COST ENGINEERING ANALYSIS

COST ENGINEERING ANALYSIS

A GUIDE TO THE ECONOMIC EVALUATION OF ENGINEERING PROJECTS

WILLIAM R. PARK, P.E.

Principal Economist
Midwest Research Institute
Kansas City, Missouri

A WILEY-INTERSCIENCE PUBLICATION

JOHN WILEY & SONS, New York • London • Sydney • Toronto

Library of Congress Cataloging in Publication Data:

Park, William R.
Cost engineering analysis.

"A Wiley-Interscience publication."
Includes bibliographical references.
1. Engineering economy. I. Title.

TA177.4.P37 658.1'554 72-10237
ISBN 0-471-65914-2

Printed in the United States of America

10 9 8 7 6 5 4

Preface

The main purpose of *Cost Engineering Analysis* is to insure that technically feasible engineering projects will also be economically attractive business ventures.

Good engineering involves business judgment in addition to engineering judgment. More than ever before, engineers must contribute to management decisions regarding the allocation of company funds to achieve the greatest possible benefits, whether in terms of profitable projects, salable products, or marketable services. Engineering without economics is a poor excuse for engineering and is certainly bad business. Maintaining the financial strength of his company is as important a responsibility of the engineer as maintaining its technical strength. Market comprehension and a knowledge of economics, along with the ability to work closely with market specialists, accounting personnel, customers, and others, are important qualifications for a competent engineer.

It is my hope that *Cost Engineering Analysis* will accomplish at least three important objectives:

1. Provide practicing engineers with the necessary economic tools for analyzing projects and for preparing sound, easily understood, investment proposals.
2. Provide management with the economic background necessary to thoroughly understand and properly evaluate engineering investment proposals, and to help establish a more objective and realistic basis for decision-making.
3. Promote a better understanding among engineers, accountants, and managers by demonstrating the close economic relationships between these

important disciplines, and by showing how the same basic economic principles apply to all.

This volume presents the basic tools of economic analysis necessary for evaluating engineering-based proposals. It shows how sound economic principles, applied to the analysis of engineering projects, can lead to the best possible decisions by management. With equal importance it shows how neglect of these basic principles can lead to costly, uneconomic, misleading, and sometimes financially disastrous results. Throughout, *Cost Engineering Analysis* stresses the practical applications of economic analysis from the engineer's viewpoint, using numerous examples to illustrate these applications.

The first two chapters present the necessary background information regarding the functions and objectives of engineering economic analysis, the steps involved in making feasibility studies, and the essentials of preparing and presenting a feasibility report. The time value of money is explored in considerable depth, including the use of interest formulas and tables and the outside factors that influence interest rates.

The next three chapters deal with the different approaches to investment analysis and the means by which investment performance can be measured. Some of the topics included in these chapters are the reasoning behind investment decisions; various criteria for evaluating investment alternatives; different ways of calculating return on investment, including short-cut graphical techniques that can save many hours of laborious calculation time; and the application and administration of postinvestment analyses to increase the effectiveness of capital expenditure programs.

The following five chapters deal with different kinds of costs and how they can be handled. They include depreciation accounting and estimation of economic project life; determination of the cost of equity and debt capital; classification and analysis of various types of costs; estimation of capital and operating costs even with limited data, along with estimation of the accuracy of the estimates; and equivalent annual costs as a means of comparing alternatives on a consistent and comparable basis.

The next two chapters cover breakeven and profit analysis. Here the effect of variables such as overhead costs, percentage markups, and sales volumes are considered as they bring about changes in a company's or project's breakeven and profit structure. Important financial ratios which can be readily developed from a firm's balance sheet and operating statement, and which, when properly interpreted, can lead to valuable insights regarding the efficiency of the company's use of its financial resources, are also discussed.

The final three chapters disclose some special tools and techniques of the professional engineer-economist used in business and economic forecasting, risk evaluation, and profitability modeling. Sources of economic data are identified and basic forecasting techniques are presented; methods for measuring and evaluating risk are described, along with techniques for minimizing the adverse effects of these risks and for relating profitability goals to the risk level; and the development of broad-scope economic models is discussed, where everything fits together—markets, prices, costs and capital requirements—into cash flow and profitability models that effectively and accurately describe the overall economic environment and conditions under which a company can expect to operate in an otherwise uncertain future.

Much of the material presented in this book was originally published in the "Engineering Economics" column of *Consulting Engineer*. The enthusiastic reception of this column since its January 1966 beginning, coupled with numerous requests from consulting engineers to make the material available in a more permanent and accessible form, provided the main incentive for reorganizing and assembling the material into this book. The cooperation of *Consulting Engineer*, and especially its editor, Arthur Steinmetz, is genuinely appreciated and gratefully acknowledged.

Others to whom special thanks are due for a variety of reasons include many associates at Midwest Research Institute, including Dr. Charles N. Kimball, President; John McKelvey, Vice President and Director of Economics and Management Science; Gary Nuss, Manager of Technoeconomic Programs; and J. B. Maillie, Principal Economist. Finally, Peggy Allensworth, who prepared the manuscript, merits special gratitude.

WILLIAM R. PARK

Kansas City, Missouri
August 1972

Contents

COST ENGINEERING ANALYSIS

1

Introduction to Feasibility Analysis

Engineers are frequently called upon either to pass judgment on projects involving substantial expenditures, or to submit informed proposals on these investments for management consideration. With the trend toward automation—resulting in larger initial investments and correspondingly large fixed or capital costs—economically sound investment decisions are becoming increasingly important. Economic analysis now plays an essential role in virtually every engineering project and management decision.

THE ELEMENTS OF ENGINEERING-ECONOMIC ANALYSIS

Good management consists primarily of making wise decisions; wise decisions in turn involve making a choice between alternatives. Engineering considerations determine the *possibility* of a project being carried out and identify the alternative ways in which the project *could* be handled. But economic considerations largely determine the project's *desirability* and tell whether or not it *should* be done and, if so, where, when, and how.

Feasibility studies constitute an extremely important part of most engineering projects. A feasibility study determines either the *which* or the *whether* of the proposed project: *which* way to do it, or *whether* to do it at all.

The term feasible means simply "capable of being dealt with successfully." In an engineering sense, then, feasibility means that the project being considered is technically possible—that it *can* be carried out. Economic feasibility, in addition to acknowledging the technical *possibility* of the project being carried out, further implies that it can be justified on an *economic* basis as well. Economic feasibility measures the overall desirability of the project in financial terms and indicates the superiority of a single approach over others that may be equally feasible in a technical sense.

OBJECTIVES OF FEASIBILITY ANALYSIS

The main objective of a conventional economic feasibility study is to identify and evaluate the economic outcome of a proposed project so that whatever funds are available can be used to the best (or at least to good) advantage. An engineer's economic feasibility study is frequently used as the basis for obtaining funds for financing public works projects and as the basis for allocating or appropriating funds by industrial firms. Used in this way, the engineer's feasibility report must provide, in a readily understandable form, all data needed by top management to reach a sound economic decision regarding the disposition of large sums of money.

Regardless of the type of project being considered, the feasibility study invariably requires that estimates be made of cash flows and economic benefits over some period of time in the future. For public works programs the time period may correspond to the economic life of the project, while in other cases a private company may specify a different—probably shorter—study period.

In any event the feasibility study is always made from the viewpoint of whoever is spending the money and usually involves a comparison of alternatives on a monetary basis. Even when only one approach to a problem is being considered, alternatives still exist; in this case the comparison is between the alternatives of doing it or not doing it.

PROCEDURE FOR FEASIBILITY STUDIES

Most engineers can recall the "scientific method," which involves five distinct phases: (1) observation, (2) problem definition, (3) formulation of hypothesis,

(4) experimentation, and (5) verification. A similar sequence of nine clearly defined steps is involved in making a thorough economic analysis:

1. Understand the problem.
2. Define the objectives.
3. Collect the data.
4. Interpret the data.
5. Devise alternative solutions.
6. Evaluate the alternatives.
7. Identify the best alternative.
8. Implement the best alternative.
9. Monitor the results.

The first seven steps might be considered the "feasibility analysis" phase. Actually, steps 2 through 7 make up the conventional feasibility study for most outside clients. The client company identifies its problem (or at least is aware that a problem exists) before it requests a feasibility study. Then, on completion of the study, the company carries out the last two steps, implementing the recommended plans and monitoring their results. In preparing investment proposals for his own company, though, an engineer who prepares the initial study may also have to live with the results of his recommendations.

While the same general sequence of steps is involved in nearly all feasibility studies and economic analyses—in all problem-solving situations, in fact—many of the steps overlap and can be carried on concurrently. Figure 1 shows a general time sequence chart for typical feasibility studies.

UNDERSTAND THE PROBLEM. Much time has been wasted solving problems that never needed to be solved. Making a mistake in the first step in solving *any* problem is bad enough, but when the first step is to identify the problem, a mistake is disastrous. Nevertheless, it is in defining the problem that many feasibility studies fall short. There is often a substantial difference between what a company *thinks* its problem is and what its problem *really* is. While most problems can be solved, avoided, or lived with, there are several observations that should be made here regarding problems in general:

1. Do not attempt to solve a problem without first making sure there is one.

2. If there is a problem, avoid putting a lot of effort into an attempt to solve it unless there is some reason to believe that it *can* be solved.

3. If there *is* a problem that *can* be solved, be certain that there is some benefit to be gained by its solution before spending too much time working on it.

4. Solve important problems first. There are plenty of big problems to worry about without wasting time on little ones.

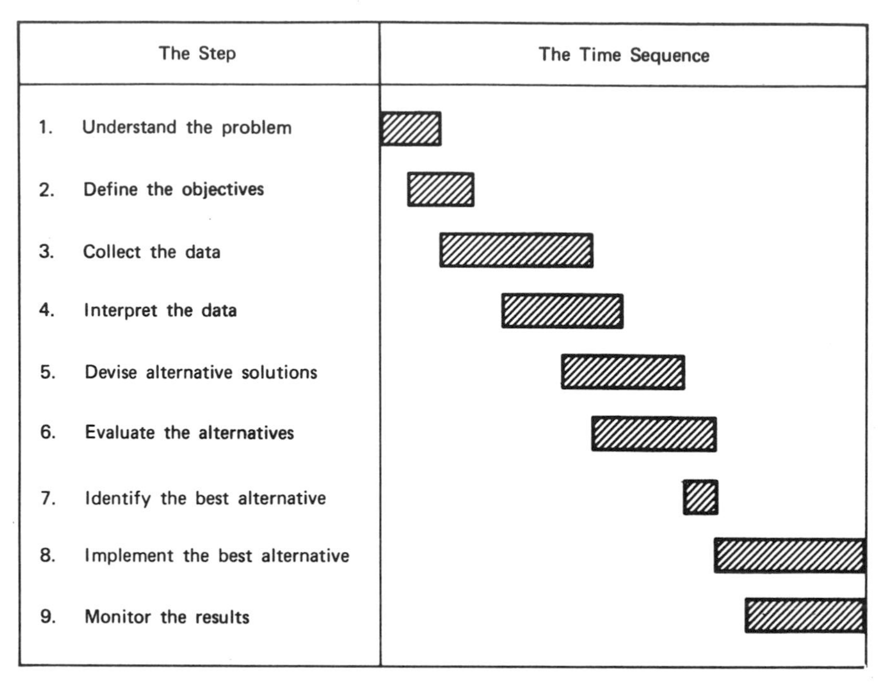

Fig. 1. *Sequence chart for engineering economic analysis.*

DEFINE THE OBJECTIVES. Failure to clarify the objectives of a feasibility study has probably resulted in more unsuccessful projects and dissatisfied clients than any other single factor. Definition of a firm's objectives, then, should be made at the outset—perhaps even before the problems have been completely analyzed. A company might have many different objectives in beginning a new project. This is perfectly acceptable so long as the objectives are clearly stated and compatible with each other. Some typical ojectives might be to:

1. Achieve a specified production or sales volume.
2. Attain a specified gross profit.
3. Earn a specified percentage return on sales.
4. Earn a specified percentage return on total assets.
5. Earn a specified percentage return on invested capital.
6. Improve company performance in a specified manner.
7. Obtain a specified share of the market.
8. Solve a specified operating problem.
9. Pay out within a specified time period.

Regardless of the firm's objectives in conducting a feasibility study, the important point in all cases is that the objectives be *specified* in some quantitative, measurable terms, and that the procedures for measurement be clearly stated. If the feasibility study is to determine *whether* or not to become involved in a proposed project, then the specified objective will provide the cutoff point; if the proposed plan can satisfactorily accomplish the objective then the plan will be approved. Or, if the study is to identify *which* of several different ways is the *best* way of accomplishing something, then the relative desirability of alternative plans must be measured in terms of how well they meet the objective criteria.

COLLECT THE DATA. Data collection can begin as soon as the problems and objectives are well enough defined to indicate the types of data that will prove useful in carrying out the study. All data pertaining to the company's specific problems and objectives should then be collected and maintained in current order. Much of the data will be of a historical nature, and a good backlog of historical economic data can prove invaluable to the engineer involved in these types of studies.

The data collection phase should normally begin with a review of published literature on whatever subjects are being investigated; too much time has been wasted in trying to develop data that were already available to those who knew where to look. Whatever data are not readily available from the literature can frequently be obtained through private company sources; otherwise, substitutes for data must be improvised through assumptions and estimates. As data are collected, they should be carefully referenced so that original documents can be retrieved if needed. Also, a clear distinction should be made between factual data, estimates, and assumptions. The reliability of the conclusions drawn from data is no better than the reliability of the data themselves.

In the collection of data, economic studies—unlike technical studies—can seldom employ experimentation to develop new data. Economic factors are generally less controllable than physical conditions, and so many economic variables may be involved that their interrelationships are unclear.

INTERPRET THE DATA. Interpretation of the data collected in the preceding step must necessarily wait until most of the data have been collected. Raw data should be organized in such a way that each piece of information, complete with references, can be found when needed. Proper organization of data makes their analysis much easier.

Most data are quantitative and well adapted to statistical analysis. But the analysis task can often become a burden, and important information and relationships may remain hidden because of the sheer bulk of the data. Consideration should be given to the use of a computer to aid in data evaluation. A computer can often extract the maximum amount of information from raw data and make the whole analysis task more manageable, at minimum cost.

The whole objective of data organization and analysis is simply to help make some sense out of the mass of data that has been collected—specifically, to interpret the data into meaningful terms that will help management make the best possible decisions. Data analysis is in fact one of the most fascinating parts of the whole study. While the data do not necessarily indicate what can be expected in the future, they at least tell what has happened in the past, and why. What has happened in the past usually provides a good indication of what the future holds. The data provide a sound and objective basis for gaining a knowledge of the past and an understanding of the present.

DEVISE ALTERNATIVE SOLUTIONS. Alternative plans may be suggested while the data interpretation phase is still in progress. Identifying alternative methods of satisfying an established set of physical requirements is essentially an engineering problem and may have little regard for the suspected economic merits of the proposed alternatives. Given the data to work with—and assuming a firm understanding of all that the data imply—this step involves some creative thought. It is unlikely that any single clear-cut plan can be devised that will accomplish all the firm's objectives under the constraints imposed by the economic environment. Compromises and tradeoffs will undoubtedly be required later, but the important point at this time is to not igonore any possibilities that might later prove worthy of consideration. If the project involves producing and selling a commodity or a product, then different manufacturing methods and different levels and patterns of sales should be considered. If the project is a manufacturing cost study, then the possibility of changes in labor costs or productivity, interest rates, taxes, and other cost factors can be introduced at this time. Any factor that appears at all reasonable should be proposed for evaluation.

EVALUATE THE ALTERNATIVES. The expected results of alternative proposals can be measured as they are developed. Here it is important to have a clear understanding of exactly *how* the alternatives are to be evaluated. If a percentage-return-on-investment objective has been established in the second step, then the method of calculating return on investment must also be specified. To some, return on investment implies a discounted cash flow analysis; to others, the reciprocal of the payout time might mean the same thing.

The economic desirability of a project depends primarily upon the magnitude and timing of cash flows, weighed in light of previously established criteria. When the approaches have been clarified, this step is essentially arithmetic, applying the prescribed analysis techniques (discounted cash flow, payout period, equivalent annual costs, etc.) to the data already collected. If the company has established a cutoff point in terms of return on investment or payout time, then the alternatives failing to reach the cutoff point may be eliminated from further consideration. Otherwise, the alternatives may simply be ranked in order of desirability. But even the rejected alternatives should be documented so that, should the need arise, their initial consideration can be shown and the reasons for their rejection justified.

SELECT THE BEST ALTERNATIVE. Selection of the best alternative must wait until all the alternatives have been evaluated in the preceding step. Sometimes the choice is the responsibility of the persons who are conducting the feasibility study; more likely, the choice is a top-management decision in which selection is made from among the best few choices. Probably, most of the alternatives originally developed in the fifth step will not appear to be particularly desirable after they have been objectively evaluated, and the choice will be narrowed down to just a few "best" alternatives. From this point the decision regarding which course of action to follow may be extremely difficult. But at this point, presumably, the only remaining choices are among the relatively good ones, so the worst that can happen is that a good—although not necessarily the best—choice is made. An *acceptable* choice, though, is almost a certainty if the analysis has proceeded as outlined here.

Selecting the best alternative requires identifying the one that comes closest to accomplishing the firm's desired objectives. There may be cases in which *none* of the alternatives meets the objectives; this indicates either a poor choice of alternatives or a poor choice of objectives, and one or the other must be changed. But one plan is bound to come *closest* to accomplishing the objectives, and if action is required this plan offers the best *available* choice.

Since the "golden rule" of investment economics is to "maximize present discounted value," the most economically desirable alternative is the one offering the highest rate of return—providing this is the criterion established by the company. Some companies may impose unreasonable demands on a new project by requiring an unrealistically high rate of return or an unreasonably short payout period. However, the main purpose of a feasibility study is to *analyze* and to *present* the *results* of the analysis, not necessarily to *recommend* a particular course of action. The feasibility study should objectively and quantitatively show the economic results and implications of following different actions; whether or not further action is taken may depend upon factors and policies beyond the engineer's knowledge, control, or responsibility.

IMPLEMENT THE BEST ALTERNATIVE. As soon as evaluation is complete and the best alternative (or alternatives) has been identified, the feasibility study goes into administrative channels. The engineer's role in this phase may ultimately range from complete responsibility for design and construction to almost complete inactivity. In either case those who prepared the study and presented the plans for management consideration should participate in the implementation phase at least enough to insure that the plans that were proposed are the plans that are being implemented. Also, any serious deviations from the original estimates should be quickly noted, and their impact on the overall attractiveness of the project should be appraised and reported.

MONITOR THE RESULTS. The moment of truth arrives when the engineer's estimates are compared with the actual results of having acted on his analysis and findings. The effectiveness of any proposal can be measured only in terms of how well it accomplishes the objectives. If the objectives and goals have been adequately defined, the measurement is a relatively simple task. Some important benefits can accrue to the engineer through careful monitoring of the results of projects and comparison of the results with his estimates. Perhaps the greatest benefit is in the improvement of technique in making future estimates; another is the added care that is put into an estimate, knowing that it will later be viewed in the light of reality.

THE FEASIBILITY REPORT

The bridge between the actual performance of a feasibility study and the implementation of its findings and recommendations is the feasibility report. The main purpose of a feasibility report is to obtain some type of action or approval of the ideas presented. Regardless of how thoroughly a project has been investigated and how firmly the report's author is convinced of the project's merits (or lack of merits), the report must reach the person responsible for approval and impress *him* with the thoroughness of investigation and the reliability of the conclusions. If the report fails to reach the right person, or fails to convey the intended meaning to him, much time and effort will have been wasted.

Obtaining the desired action from the feasibility report simply requires that the report present the right ideas, accurately and as briefly as possible, to the appropriate people in a logical and impressive manner. A sloppy job of presentation implies sloppy performance in other areas, whether or not such an implication is justified. A misspelled word or typographical error can cast doubt on the accuracy of many hours of meticulous calculations.

Many good books are available that deal with engineering reports and technical writing and should be consulted by the engineer responsible for preparing reports and proposals. Recommended report formats vary in their minor details, but most can be divided into three main parts: (1) front matter, (2) body, and (3) back matter.

FRONT MATTER. The front matter includes the report cover, title page, preface, acknowledgments, table of contents, list of illustrations, list of tables, and summary. The summary is particularly important, since some readers go no further. The summary should tell briefly *what* is proposed and *why* it is worthwhile.

BODY. The body of a report includes the introduction, discussion, conclusions, recommendations, and references. This part of the report is of course the most important and most difficult part to prepare. It should be organized carefully and logically, and it must be unmistakably clear. A good outline is indispensable in structuring the report body; the outline should be prepared before the actual report writing begins. Each of the major sections in the report should, as far as possible, have approximately equal significance. A uniform system of numbering, indentation, headings, and so on should be employed throughout the report.

BACK MATTER. The back matter consists of appendixes, the index, and the back cover. The appendix can include data, formulas, and other detailed backup information which is of technical interest but not necessary for an understanding of the main body of the report.

Few reports include all these sections, although the general sequence of presentation should hold even in a letter report of a few pages.

A *letter of transmittal* may accompany the feasibility report separately, or may be bound into the front of the report just after the title page. The transmittal letter *informs* the recipient that the report is being sent, *identifies* the subject of the report, and *refers* the recipient to those individuals responsible for the report's preparation.

The accompanying exhibits (Figures 2 and 3) illustrate the general format and contents of typical economic feasibility studies. Both reports were about 40 pages in length. The study described in Figure 2 involves a feasibility analysis of a proposed total energy system for an apartment complex. Here the analysis is concerned with comparing a high-initial-cost, low-operating-cost system with a low-first-cost, high-operating-cost alternative.

Figure 3 describes a study of proposed improvements for a municipal water system. In this case some action is required and the purpose of the study is to determine the most economical alternative that adequately satisfies the physical demands placed on the system.

THE HUMAN ASPECTS OF ENGINEERING PROPOSALS

There are times when even the best ideas, presented most logically in a technically perfect form, cannot sell the proposed project; in such situations

Feasibility of a Total Energy Installation

Summary

I. Introduction
II. Comparison of alternative energy systems
III. Determinants of economic feasibility
IV. The apartment project
  A. General description
  B. The proposed energy systems
V. Analysis of energy system operations
  A. Electric loads
  B. Equipment utilization and fuel consumption
  C. Heating–cooling system fuel requirements
VI. Estimated costs and revenues
  A. Capital costs
  B. Operating costs
  C. Operating revenues
VII. Economic feasibility
  A. Present worth
  B. Return on investment
  C. Payout time
VIII. Effect of economic variables
  A. Timing of future construction
  B. Variations in occupancy rates
  C. Variations in net annual savings
IX. Conclusions and recommendations

Appendix

Fig. 2.

the human relations aspects, the intangible factors, can make the difference between acceptance and rejection of an otherwise sound engineering or economic proposal.

This broad area of human engineering is an unfamiliar one to many engineers, especially those who are accustomed to working primarily with facts, data, and problems that can be solved through objective and purely analytical methods, with little or no concern for human elements. Perhaps some engineers feel that it is outside their experience, beyond their ability, or not their responsibility to actually "sell" a project.

But whether engineer, pipefitter, bricklayer, or accountant; whether lawyer, plumber, mechanic, or professor; whether electrician, physician, preacher, or

Report on Water System Improvements

Introduction

- A. Purpose
- B. Scope of study
- C. Historical background

Summary of findings and recommendations

- I. Demand for water
  - A. Average day demand
  - B. Maximum day demand
  - C. Maximum hour demand
  - D. Summary of demand rates
- II. Existing water system
  - A. Supply
  - B. Storage
  - C. Distribution
- III. Proposed Improvements
  - A. Supply
  - B. Storage
  - C. Distribution
- IV. Estimated Costs
  - A. Short-range program
  - B. Long-range program

Appendix

Fig. 3.

peanut vendor—everyone is selling something whether he realizes it or not. He is selling his product, his service, or himself.

An engineering project is sold through the presentation of a feasibility report. Engineering services are often sold by written or oral presentation of a proposal or capabilities description. On completion of a project, the results are sold in the engineering report presented to the management or client.

The traditional method of making a technical presentation—either written or oral—is often described as consisting of three parts:

1. Tell the audience what you're going to say.
2. Say it.
3. Then tell the audience what you said.

This approach is satisfactory as far as it goes and can be adequate for many written technical reports. However, if the objective of a presentation is to sell a proposed project to people, then the people must somehow be brought into the presentation.

People are the most neglected part of a typical engineering proposal or report. But until people are considered an essential part of the presentation, neither the project nor the engineer will be salable. The reader or listener wants to know not only about the project itself, but what the project can do for him or his company.

One of the most popular sales presentation techniques—dating back at least to the 1920s—is called the AIDA approach, with the initials standing for *A*ttention, *I*nterest, *D*esire, *A*ction. By following this approach an effective presentation to an audience, either a group of people or an individual, can accomplish these four objectives:

1. Attract *attention* to the presentation.
2. Arouse *interest* in the subject.
3. Generate a *desire* for the project, product, or service.
4. Motivate the desired *action.*

The last three steps in the AIDA presentation can be related to the previously mentioned three-step approach to technical reporting. *Interest* can be aroused by briefly describing and emphasizing the most important (and therefore the most salable) points to be covered in the presentation. *Desire* for the project should evolve naturally during the listening to or reading of the presentation itself, as facts are first presented, then interpreted, finally evolving into meaningful and relevant conclusions. *Action* results from translating the conclusions into positive recommendations for what needs to be done.

It is the *attention* phase that is often left out of the engineering proposal—where the person or organization authoring the report or making the presentation must establish a clear relationship between himself, the subject of the presentation, and the audience to which the presentation is directed. Specifically, he owes his audience some justification for

- being there, either in person or represented by a written document,
- presenting this subject,
- to this audience,
- at this time.

The audience's attention is earned only if these points are adequately clarified. The first 30 seconds or 50 words of either a written or oral presentation must command the initial favorable attention that will make the reader or listener *want* the presenter to continue his story.

The first 50 words in a presentation are in fact more important than the next 10,000 words in terms of their overall impact on the audience. There is no second chance to create a first impression.

A good "human engineer" knows that if he is unable to establish a firm relationship between himself, the subject, and the audience within the first 30 seconds of his presentation the audience may be lost forever—mentally, if not physically. It is these 30 seconds or 50 words that either provide a good springboard from which to jump into the formal presentation, or impose a handicap upon the rest of the presentation from which it may never recover.

SUMMARY

The increasing capital costs associated with engineering projects have increased the necessity for sound engineering-economic analysis. An economic feasibility study determines either *which* of several different ways a project should be carried out, or *whether* the project should be carried out at all. The main objective of the feasibility study is to predict the outcome of a proposed expenditure in financial terms so that available funds can be put to their most advantageous use. There are nine steps involved in a complete engineering-economic analysis: (1) understand the problem, (2) define the objectives, (3) collect the data, (4) interpret the data, (5) devise alternative solutions, (6) evaluate the alternatives, (7) identify the best alternatives, (8) implement the best alternatives, and (9) monitor the results. The first seven steps make up the usual feasibility study, and when they have been completed a feasibility report is generally required so that appropriate management action can be taken to initiate the project. The feasibility report should summarize the important features of the proposed project in a brief but impressive presentation.

2

The Time Value of Money

Since most business ventures involve the use of other people's money, interest —the rental charged for the use of borrowed money—plays an important role in the engineer's preinvestment analysis. From a borrower's standpoint the opportunity to invest borrowed money at a higher rate than must be paid for its use justifies the payment of interest. From the lender's standpoint, interest represents his compensation for not being able to spend the money elsewhere. These, then, are the two primary reasons for having interest: the opportunity to invest money and the desire to spend it.

FINANCIAL MARKETS

Financial markets can be broadly classified into two categories: (1) the capital, or investment, market and (2) the money market. While the distinction between capital and money markets is not always clear because of some overlapping, the length of time for which funds are committed is probably the best way to distinguish between them. The money market usually refers to funds borrowed or loaned for a year or less, while the capital market encompasses longer-term obligations. Overall, savings institutions provide about half of the total funds available in the financial markets, while commercial banks provide about 30 percent and the remainder is supplied by business corporations and individuals.

THE CAPITAL MARKET

The capital market, where most investment-type funds are placed, is made up primarily of five types of obligations:

1. Corporate Bonds.
2. United States government bonds.
3. State and local bonds.
4. Corporate equities.
5. Mortgages.

In the capital market mortgages account for more than half of all funds used, with the rest divided between corporate and government securities. More than two-thirds of the funds employed in the capital market are obtained from savings institutions.

Bonds are long-term obligations of a specific value offering assured interest at an established rate. They may be issued by corporations, states, muncipalities, politically established districts, or the federal government. Bonds, to be attractive to investors, must be heavily secured, whether by assets, anticipated revenues, or other means. Bond repayment schedules are usually set up to pay interest periodically—one to four times a year—with the principal amount payable upon the bond's maturity. Most bonds are sold on the primary market (when newly issued), although there is also a large secondary, or resale, market for many of them.

This situation is just the opposite of the *equity* market, which has a small primary sale on new issues but a huge secondary market. Thus the equity market makes up a relatively small portion of the total funds available for corporate financing. Corporate equities, or capital stocks, may be issued by firms whose cost of capital would be lower than could be obtained by entering the bond market, or whose financial position is such that there would be insufficient market for their bonds.

Mortgages are long-term securities given for repayment of a debt usually connected with real property. The mortgage is secured by title or lien on the property being financed. Repayment of a mortgage is usually made in a series of uniform amounts, with each payment including both principal and interest.

THE MONEY MARKET

The money market—the shopping center for short-term funds—includes these major segments:

1. Treasury bills.
2. Commercial bank loans.
3. Commercial paper.
4. Bankers' acceptances.
5. Certificates of deposit.

Just over half of the short term funds are supplied by commercial banks, with about 30 percent coming from business corporations and the remainder from insurance companies, credit unions, other investor groups, and individuals. The most important single use of the available short-term funds is in the field of commercial bank loans.

Treasury bills—short-term notes issued monthly by the U.S. Treasury Department—provide funds needed for the federal government and offer excellent short-term investments for banks, business corporations, and individuals. Since treasury bills are sold through competitive bidding, their interest rates are probably the most sensitive in the money market to changes in the economy.

Commercial bank loans are made to established corporate customers at a negotiated rate of interest, "prime" or above. (The prime rate refers to the interest rate that commercial banks charge their preferred customers for short-term funds). They may be secured or unsecured, depending on the customer's credit standing and financial situation, and are issued for varying lengths of time.

Commercial paper refers to secured or unsecured promissory notes issued by corporations and sold to investors, most of whom are other corporations. These notes are written, unconditional promises to pay—on demand or after a short time—a specified amount of money. The notes are often discounted at commercial banks.

Bankers' acceptances are bills used to finance the import, export, transfer, or storage of goods. The bank guarantees their payment at maturity.

Certificates of deposit (CDs) refer to money deposited in commercial banks for a specified time, commonly 3 months, drawing interest at a predetermined rate. Certificates of deposit usually make up a large part of the funds available for commercial bank loans, although the situation may change should higher interest rates be obtainable elsewhere. Since the Federal Reserve places an upper limit on the interest that can be paid by commercial banks on CDs, corporations with idle funds place their money elsewhere if the rates are not competitive with what can be obtained on treasury bills or other short-term investment opportunities.

HOW INTEREST RATES ARE DETERMINED

Each type of loan—whether a mortgage, corporate bond, personal loan, or whatever—carries a different price, or interest rate. This interest rate, regardless of the type of loan involved, is a function of the supply of, and the demand for, money.

When funds are in short supply relative to demand, short-term interest rates can be expected to rise. When short-term rates go up, long-term rates cannot help but be affected. While short-term rates are determined by current supply-and-demand factors, long-term rates must anticipate supply-and-demand relationships over the future life of the interest-bearing security.

The cost of money to commercial banks naturally has a strong impact on the price they charge borrowers for the use of money. With an upper limit on the interest that banks are allowed to pay on time deposits (savings accounts and certificates of deposit), large depositors shift their reserve funds to commercial paper, treasury bills, and other high-yield, short-term securities whenever warranted by the yield differential. This in turn leaves commercial banks with less money available to lend.

But banks make money only by lending or investing other people's money; they must therefore look to other sources for their funds if they are to have money available for loans, even if the other sources are relatively expensive.

These outside funds may come from other banks, either from United States banks having excess reserves or from European banks holding United States dollars. Regardless of the source, the borrowing bank will probably have to pay a substantial premium for its outside funds.

Nevertheless, during inflationary periods borrowers are willing to pay almost any amount to obtain money, since they expect costs to go up and can probably raise prices to cover the increased cost of borrowed money. At the same time, lenders must ask high interest rates to protect themselves from a loss in purchasing power.

GOVERNMENT POLICY

Government policy is an important determinant of the rate of interest, and the direction of the nation's economy can be controlled to some extent by how the government manipulates the money market.

The Federal Reserve has the power to: (1) change its reserve requirements to member banks, thus changing the amount of money available for loans; (2) change its pattern of open-market operations, which it frequently does; and (3) change the discount rate, thereby lowering or raising the cost of money to banks. The Federal Reserve can also set an upper limit on the interest rates that can be paid on time deposits by banks under its juridisiction.

Even with these controls at its disposal, the money market sometimes gets away from the government, resulting in threats of wage and price controls and other strict measures to bring the economy back in line. Federal monetary policy, while important, is but one of many factors that influence the cost of money.

THE EFFECTS OF INTEREST RATES

High interest rates reflect a tight money supply. The resulting difficulty in obtaining money, or the high price that must be paid for it when it is obtained, show up in the nation's economy in several ways.

First, high interest rates make it more expensive to make a new investment. This is especially important for public utilities and municipalities, whose funds must be raised from outside sources through the issuance of bonds. Such fixed-income, long-term securities are generally unattractive during periods of rising prices, as evidenced by the difficulties experienced by state and local governments in marketing bonds to finance schools, roads, sewers, and other public improvements.

A second major effect of high interest rates is that, even with internally generated funds, new capital projects appear less desirable relative to the returns offered by alternative short-term investments such as U.S. Treasury bills. When capital investments must be financed with borrowed money, high interest rates reduce the effective rate of return on invested capital. If the expected profitability of a proposed investment does not compare favorably with the pervailing interest rate—a convenient standard for comparison—considering such factors as risk and liquidity, a firm will not make an investment.

In the private construction field, where mortgage financing is commonly employed, interest rates also have a strong effect. Doubling the rate of interest on a mortgage more than doubles the amount of interest paid. Raising the interest rate on a 20-year mortgage from 5 to 10 percent, for example, actually increases the total amount of interest paid by about 2.24 times; a 20-percent increase in the interest rate increases interest payments by about 23 percent.

Thus slowdowns in home-building, delays in public projects, and decreased capital outlays by large corporations all may result, directly or indirectly, from high interest rates. A general feeling of uncertainty about the immediate future of the money market also contributes heavily to an overall slowdown in the economy.

INTEREST FORMULAS

There are but three basic interest formulas that need concern the engineer. These, with their reciprocals, permit calculation of the following six important factors. In all these formulas, i represents the rate of interest, usually expressed on an annual basis; n indicates the number of interest periods; P is a present amount of money, or the equivalent present value of some future amount; S is the future amount of money, either payable or receivable in a single lump sum; and R is a uniform annual amount.

Figures 4–6 show the factors pertaining to each of the six compound interest formulas for selected interest rates and over different time periods.

SINGLE PAYMENT COMPOUND AMOUNT FACTOR. This is the future amount S that some present amount P will accumulate to in n years at i-percent interest.

$$S = P(1 + i)^n \tag{1}$$

The value of S is plotted in the top part of Figure 4.

SINGLE PAYMENT PRESENT VALUE FACTOR. This factor is the amount P that a future amount S, receivable in n years, is now worth, with interest at i percent. This is the reciprocal of formula (1).

$$P = S\left(\frac{1}{(1 + i)^n}\right) \tag{2}$$

P is shown graphically in the bottom part of Figure 4.

ANNUITY COMPOUND AMOUNT FACTOR. This is the amount S that an equal annual payment R will accumulate to in n years at i-percent interest.

$$S = R\left(\frac{(1 + i)^n - 1}{i}\right) \tag{3}$$

S values are plotted in the top part of Figure 5.

SINKING FUND FACTOR. This factor is the equal amount R that must be invested at i percent in order to accumulate to some specified future amount

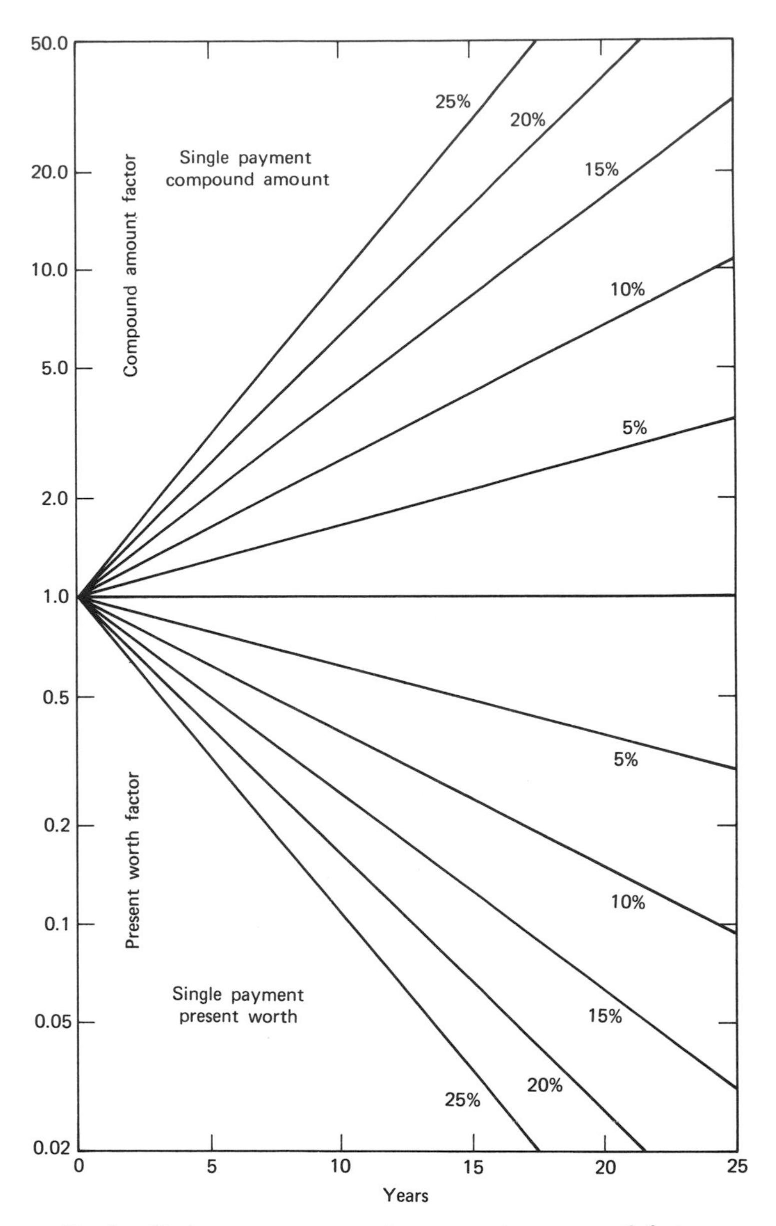

Fig. 4. *Single payment compound amount and present worth factors.*

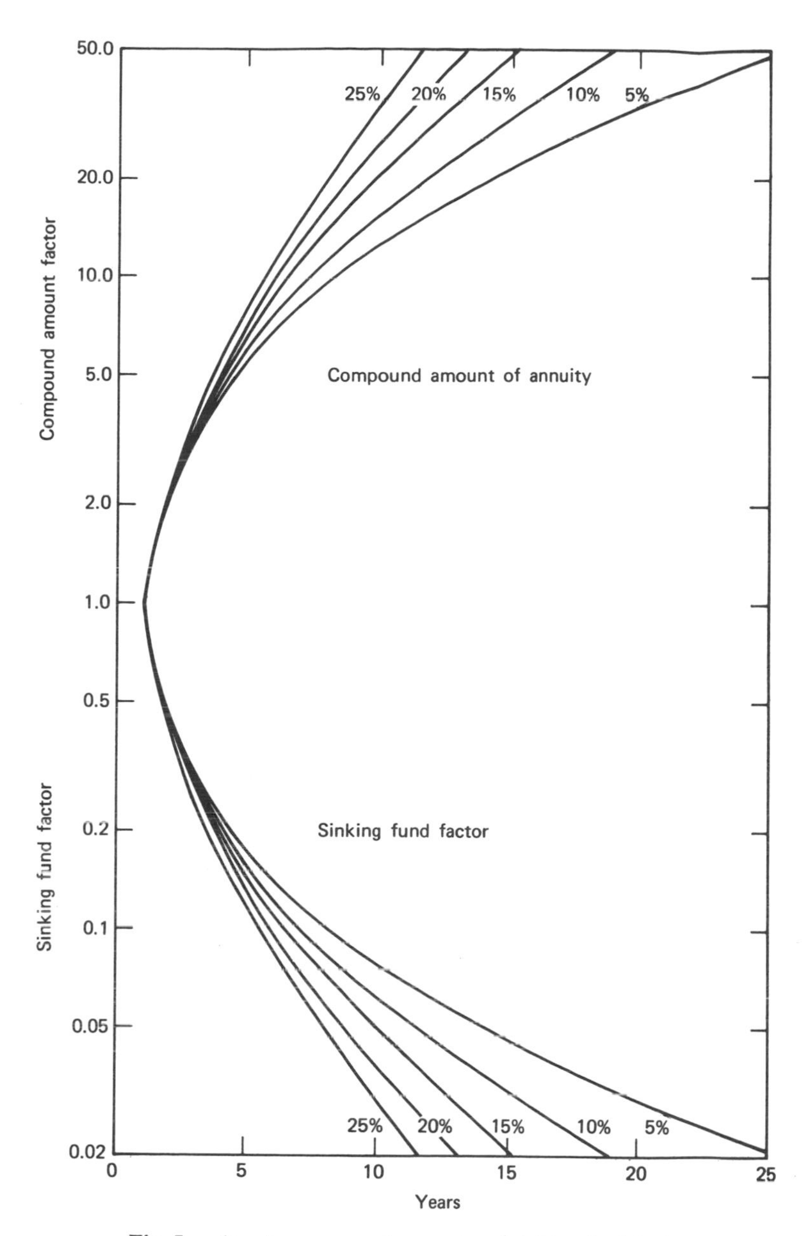

Fig. 5. *Annuity compound amount and sinking fund factors.*

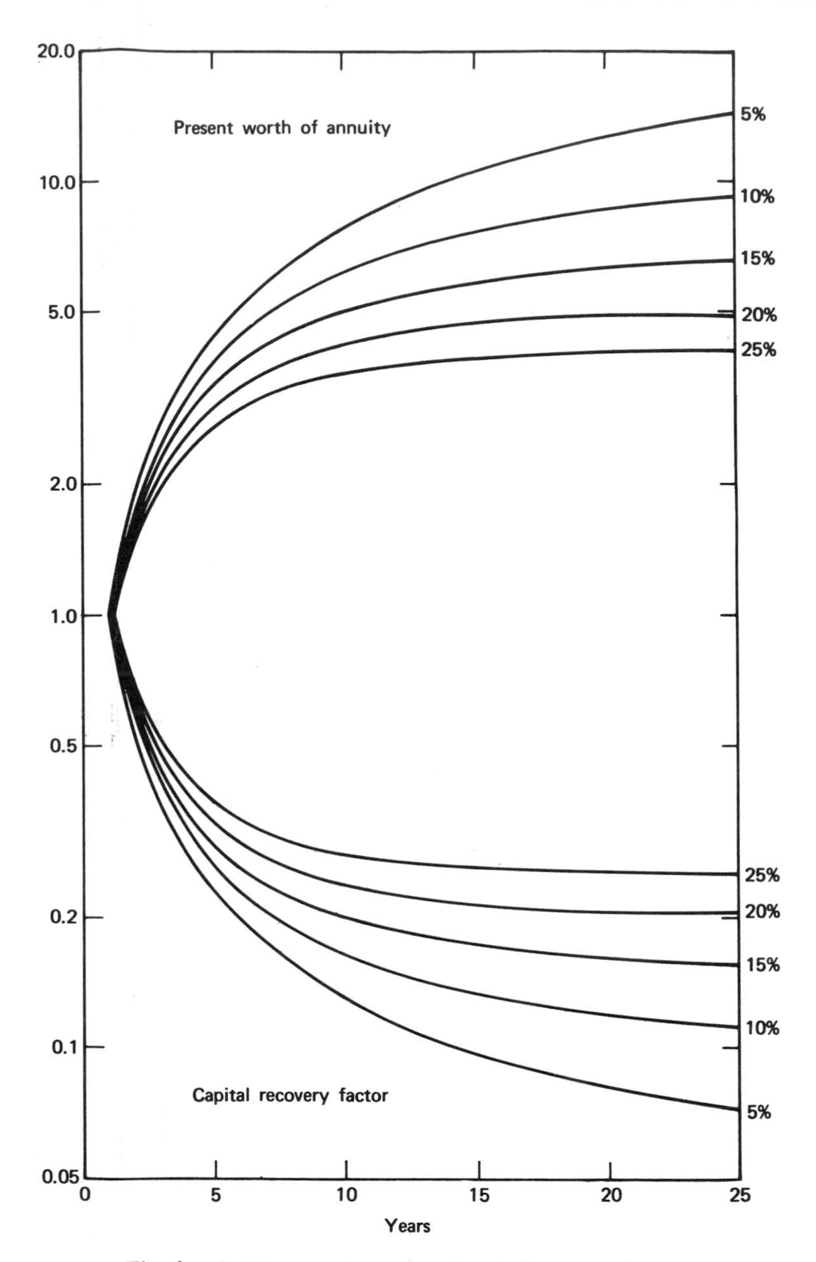

Fig. 6. *Annuity present worth and capital recovery factors.*

S over a period of n years. This is the reciprocal of formula (3).

$$R = S\left(\frac{i}{(1+i)^n - 1}\right) \tag{4}$$

R values are shown in the lower section of Figure 5.

CAPITAL RECOVERY FACTOR. This is the annual payment R required to amortize, or completely pay off, some present amount P over n years at i-percent interest. The capital recovery factor is equal to the sinking fund factor plus the interest rate.

$$R = P\left(\frac{i(1+i)^n}{(1+i)^n - 1}\right) \tag{5}$$

The bottom part of Figure 6 shows the capital recovery factors.

ANNUITY PRESENT VALUE FACTOR. This is the present amount P that can be paid off by equal annual payments of R over n years at i-percent interest; or, the present value P of an n-year annuity R discounted at i percent. This is the reciprocal of formula (5).

$$P = R\left(\frac{(1+i)^n - 1}{i(1+i)^n}\right) \tag{6}$$

These P values are shown in the top half of Figure 6.

USING THE INTEREST FORMULAS

The primary uses of the interest formulas in calculating dollar values are apparent from the preceding descriptions. But these formulas also apply equally well to anything exhibiting a constant *rate* of growth (as opposed to constant arithmetic increases). Population growth is a good example of a fairly constant rate of increase, as are per capita usage of electrical energy or water and various measures of production and consumption.

The single payment compound amount factor, for example, is the basic formula used in measuring growth rates. If the population of an area has grown from 1 million persons in 1950 to 1.7 million in 1970, these figures substituted in formula (1) indicate an average compound rate of growth of 2.7 percent annually over the 20 year period. The same formula can then be used to project the present population to any future date, assuming the rate of growth to remain constant. The compound amount factor plots as a straight line on semilog graph paper.

The four annuity-type interest formulas can be used whenever a uniform stream of receipts or payments is involved—such as in comparing a series of operating savings from a new piece of equipment, a new process, a different design, or an alternative energy source. These formulas measure, in different ways, the cumulative effect of future cash flows and allow these cash flows to be expressed as a single equivalent amount.

The capital recovery factor is especially useful in engineering economy studies in which alternatives having different useful service lives are being compared. The capital recovery factor amortizes the initial investment over its life, thus reducing the alternatives to their equivalent annual costs. It considers both depreciation (principal) and interest, thus representing direct out-of-pocket costs of a project financed entirely through outside funds.

USING INTEREST TABLES

A good set of interest tables, giving numerical values for all six of the basic interest formulas for different interest rates and time periods, is an important part of every engineering economist's library. These tables are included in most standard handbooks of mathematical tables, although their usefulness is sometimes restricted because of the limited range of values presented. In feasibility studies and investment analyses, interest rates up to 25 percent, and sometimes higher, are often used.

Table 1 shows a page from a typical interest table. By using the 10-percent compound interest factors shown in the table, solutions to all typical interest problems at this rate can be quickly found.

In the first column the *compound amount* factor shows the amount that a dollar accumulates to over N time periods. One dollar invested now at 10 percent accumulates to \$2.59 over a 10-year period, and to \$6.73 in 20 years.

The *present worth* factors in the next column are reciprocals of the compound amount factors; \$1.00 receivable in 10 years is worth only \$0.386 now; \$2.59 receivable in 10 years has a present worth of \$2.59 × 0.386, or \$1.00.

The *sinking fund* factor in the third column shows the amount that must be invested at 10 percent each year to accumulate to \$1.00 in n years. Thus \$62.75 must be invested annually to accumulate, with interest, to \$1000 in 10 years.

In the next column the *capital recovery* factor shows the annual payment required to cover principal and interest in equal annual amounts over an n-year period. A \$1000 loan can be retired in 5 years by paying back \$263.80 annually. A total of \$1319.00 would be repaid, of which \$1000 is the principal and \$319 is interest.

The *compound amount* factor in the fifth column shows the total accumulation, with interest, of an equal amount invested each period for *n* periods. If $1000 were invested each year at 10 percent for 20 years, the total amount that would accumulate at the end of the 20th year would be $1000 × 57.275, or $57,275. In this case only $20,000 represents the principal, the remaining $37,275 being interest.

The last column indicates the *present worth* of a uniform annual series of payments. This shows that to purchase a $1000, 20-year, 10-percent annuity requires a present payment of $8514. Or, a project returning $1000 annually for 20 years has a present value of $8514.

In interest tables the time periods usually are years but can be taken as quarters, months, or any other units of time. A 1.0-percent interest rate compounded monthly is *roughly* equivalent to a 12.0-percent rate compounded annually, or a 6.0-percent rate compounded semiannually, or a 3.0-percent rate compounded quarterly.

To obtain values for fractional time periods or for interest rates not included in an interest table, linear interpolation is usually adequate; if more precision is required, either a slide rule or log tables can give the answer.

ESTIMATING AVERAGE INTEREST

Since a loan is usually payed back in a series of equal payments, the outstanding balance is only about half the initial amount of the loan, so the average interest paid on the original amount is roughly half the prescribed annual interest rate. For example, if $1000 were borrowed for a year at 6 percent and paid back in monthly installments, the average outstanding balance would be about $500 (meaning that the borrower, over a year's time, actually had the use of only $500 on the average), and the average interest paid would be about 3 percent of the initial amount, or $30 total. But if $60 interest were charged on this same loan—6 percent on the initial amount—the true interest rate would be nearly 12 percent, since the $60 interest is paid on an average loan of only $500.

Some engineering economics textbooks give a formula for computing the average interest paid on the outstanding balance of a loan over its entire duration. The commonly cited "average interest" formula is:

$$\text{average interest} = \frac{i}{2}\left(\frac{n+1}{n}\right)$$

where i is the interest rate and n the number of years. This formula is actually only a rough approximation of the capital recovery factor less the straight-line depreciation rate. For example, the average interest on a 6-percent,

10.00 PERCENT COMPOUND INTEREST FACTORS

N PERIODS	SINGLE PAYMENT: COMPOUND AMOUNT FACTOR GIVEN P TO FIND S	SINGLE PAYMENT: PRESENT WORTH FACTOR GIVEN S TO FIND P	UNIFORM ANNUAL SERIES: SINKING FUND FACTOR GIVEN S TO FIND R	UNIFORM ANNUAL SERIES: CAPITAL RECOVERY FACTOR GIVEN P TO FIND R	UNIFORM ANNUAL SERIES: COMPOUND AMOUNT FACTOR GIVEN R TO FIND S	UNIFORM ANNUAL SERIES: PRESENT WORTH FACTOR GIVEN R TO FIND P	N PERIODS
	(1 + I)**N	1 / (1 + I)**N	I / ((1 + I)**N - 1)	I(1 + I)**N / ((1 + I)**N - 1)	((1 + I)**N - 1) / I	((1 + I)**N - 1) / (I(1 + I)**N)	
1	1.1000000	.9090909	1.0000000	1.1000000	1.0000000	.9090909	1
2	1.2100000	.8264463	.4761905	.5761905	2.1000000	1.7355372	2
3	1.3310000	.7513148	.3021148	.4021148	3.3100000	2.4868520	3
4	1.4641000	.6830135	.2154708	.3154708	4.6410000	3.1698654	4
5	1.6105100	.6209213	.1637975	.2637975	6.1051000	3.7907868	5
6	1.7715610	.5644739	.1296074	.2296074	7.7156100	4.3552607	6
7	1.9487171	.5131581	.1054055	.2054055	9.4871710	4.8684188	7
8	2.1435888	.4665074	.0874440	.1874440	11.4358881	5.3349262	8
9	2.3579477	.4240976	.0736405	.1736405	13.5794769	5.7590238	9
10	2.5937425	.3855433	.0627454	.1627454	15.9374246	6.1445671	10
11	2.8531167	.3504939	.0539631	.1539631	18.5311671	6.4950610	11
12	3.1384284	.3186308	.0467633	.1467633	21.3842838	6.8136918	12
13	3.4522712	.2896644	.0407785	.1407785	24.5227121	7.1033562	13
14	3.7974983	.2633313	.0357462	.1357462	27.9749834	7.3666875	14
15	4.1772482	.2393920	.0314738	.1314738	31.7724817	7.6060795	15

10.00%

16	4.5949730	.2176291	.0278165	.1278166	35.9497299	7.8237086	16
17	5.0544703	.1978447	.0246641	.1246641	40.5447028	8.0215533	17
18	5.5599173	.1798588	.0219302	.1219302	45.5991731	8.2014121	18
19	6.1159090	.1635080	.0195469	.1195469	51.1590904	8.3649201	19
20	6.7274999	.1486436	.0174595	.1174596	57.2749995	8.5135637	20
21	7.4002499	.1351306	.0156244	.1156244	64.0024994	8.6436943	21
22	8.1402749	.1228460	.0140051	.1140051	71.4027494	8.7715403	22
23	8.9543024	.1116782	.0125713	.1125718	79.5430243	8.8832184	23
24	9.8497327	.1015256	.0112993	.1112998	88.4973268	8.9847440	24
25	10.8347059	.0922960	.0101681	.1101681	98.3470594	9.0770400	25
26	11.9181765	.0839055	.0091590	.1091590	109.1817654	9.1609455	26
27	13.1099942	.0762777	.0082576	.1082576	121.0999419	9.2372232	27
28	14.4209936	.0693433	.0074510	.1074510	134.2099361	9.3065665	28
29	15.8630930	.0630394	.0067281	.1067281	148.6309297	9.3696059	29
30	17.4494023	.0573086	.0060792	.1060792	164.4940227	9.4269145	30
31	19.1943425	.0520987	.0054962	.1054962	181.9434250	9.4790132	31
32	21.1137767	.0473624	.0049717	.1049717	201.1377675	9.5263756	32
33	23.2251544	.0430568	.0044994	.1044994	222.2515442	9.5694324	33
34	25.5476699	.0391425	.0040737	.1040737	245.4766986	9.6085749	34
35	28.1024368	.0355841	.0036897	.1036897	271.0243685	9.6441590	35
36	30.9126805	.0323492	.0033431	.1033431	299.1268053	9.6765082	36
37	34.0039486	.0294083	.0030299	.1030299	330.0394859	9.7059165	37
38	37.4043434	.0267349	.0027469	.1027469	364.0434344	9.7326514	38
39	41.1447778	.0243044	.0024910	.1024910	401.4477779	9.7569558	39
40	45.2592556	.0220949	.0022594	.1022594	442.5925557	9.7790507	40

NOTE- **N IS EXPONENT N

10-year loan is approximated at 3.30 percent by using the average interest formula, while the capital recovery factor (0.1359) less the straight-line depreciation rate (0.100) sets the actual average interest rates. Use of the average interest formula therefore seldom can be justified by the engineer in any but the very crudest types of studies.

THE FREQUENCY OF COMPOUNDING

The frequency of compounding has a significant effect on the true interest rate. A carrying charge of 1.0 percent per month on an unpaid department store balance, for example, amounts to an annual rate of about 12.7 percent:

$$
\begin{aligned}
(1 + i)^n &= (1.01)^{12} \\
&= 1.127 = 12.7\%
\end{aligned}
$$

Similarly, a 1.5-percent monthly rate is equivalent to 19.7 percent paid once annually.

The expression $(1 + i)^n$ appears in all six of the basic interest formulas. If the total elapsed time is held constant (say at 1 year) and the compounding period is reduced (or, stated another way, the frequency of compounding is increased), the value of this expression increases. As the frequency of compounding approaches infinity, the expression $(1 + i)^n$ approaches e^{in}, indicating "continuous" compounding (where $e = 2.7183 \ldots$). Alternative investments all paying 6.0 percent annually, then, actually yield different amounts depending on their compounding periods. The effective interest rate on money compounded annually is 6.00 percent; semiannually, 6.09 percent; quarterly, 6.14 percent; bimonthly, 6.15 percent; monthly, 6.18 percent; and continuously, 6.19 percent.

SUMMARY

Interest represents the amount charged for the use of borrowed money. The financial marketplaces, for either invested or borrowed funds, include capital (or long-term) markets and money (or short-term) markets. The capital market is made up primarily of bonds, corporate equities, and mortgages. The money market includes treasury bills, commercial bank loans, commercial paper, bankers' acceptances, and certificates of deposit. Each type of financial obligation carries its own interest rate, determined by supply-demand relationships. The level of interest rates has a significant impact on the nation's economy, with changes in interest rates causing money to shift from one financial market to another. Probably the most important factor from a business viewpoint is the ease with which long-term capital projects can be financed. The amount of interest associated with any type of financial transaction can be calculated by using one of six standard interest formulas. These formulas are used in computing (1) the compound amount of a single payment, (2) the present value of a future payment, (3) the compound amount of an annuity, (4) the sinking fund factor, (5) the capital recovery factor, and (6) the present value of an annuity. In addition to the base interest rate, the frequency with which interest is compounded also has an important influence on the total interest charges associated with an investment.

3

Evaluating Investment Alternatives

A wide array of methods exist for evaluating capital expenditures. But the only effective way to manage investments is to think through carefully the economics of each investment proposal and to make decisions accordingly. Intuition and guesswork alone, perhaps once adequate as the basis for making capital expenditure decisions, no longer meet the pressures of competition. Making a profit from an investment has become a race against time.

There is only a limited time during which a product or process can remain profitable; maturity and declining profits are a part of every product's life cycle.

To maintain this critical period of profitable operation for as long as possible, investments and costs must be carefully controlled. With the growing trend toward automatic controls—resulting in larger capital investments —depreciation and other fixed charges against capital assets are often more significant than the direct out-of-pocket costs of labor and materials.

It is obvious that effective means of managing capital expenditures are required for financial success in any business venture. Management must avoid passing up profitable investments and avoid making unprofitable ones.

To be of maximum value, the engineer's analysis of investment feasibility must be objective, realistic, easily understandable to his client or his management, and appropriate for the situation. Often much is at stake in these studies, such as a decision to commit large sums of money, and the responsible engineer owes to his sponsor a sound knowledge of the different methods of analyzing investment proposals. Only by being knowledgeable in the tools of investment analysis can he select the approach best suited to the situation.

Basically, there are but three ways that an investment decision can be justified.

1. Degree of necessity.
2. Payout time.
3. Rate of return.

Payout time is undoubtedly the most widely used measure of the worth of an investment, although rate of return is the most logical and theoretically acceptable means of determining investment feasibility. Degree of necessity, however, still plays an important role in appropriating capital.

INVESTMENTS BASED ON NECESSITY

Capital investments based on necessity include many of the investments that an investor would rather not make but must. Typically included in this unpleasant category are:

1. Investments on equipment that breaks down and must be replaced.
2. Investments in capital assets required to meet governmental regulations, such as facilities for air and water pollution control.
3. Investments required simply to remain in business, such as the replacement of a disaster-stricken facility.
4. Investments that, even though relatively unprofitable, are necessary to meet competition.

Investments based on necessity, while not necessarily profitable, are usually more profitable than any available alternatives, since the alternatives are apt to be extreme—such as shutting down altogether. The main difficulty encountered in appraising these investments is the lack of any real objective means of measuring necessity in absolute terms.

THE PAYOUT TIME CRITERIA

The payout period—the most used measure of investment feasibility—has as its main virtue simplicity. Payout time is simply the number of years required for cash earnings or savings generated by a proposed project to equal the original capital investment. As such, the payout time serves as a simple, rough, and readily understood index of project desirability.

Payout time often is used as a screening device to identify projects that are apt to be either exceptionally profitable or unprofitable during their early years. But for most purposes payout time does not provide an adequate measure of investment worth, since many important elements of profitability are not considered. Payback, as a measure of investment desirability, has three important shortcomings:

1. It overemphasizes the importance of early cash returns in the capital expenditure program.
2. It ignores the project's economic life.
3. It fails to consider project earnings after the initial investment has been recovered.

In spite of these shortcomings, the conceptual simplicity of the payout method has certain merits in capital expenditure decisions and actually may provide an adequate measure of investment worth in some situations.

Consider, for example, the investment proposal for project A shown in Table 2.

The investment being considered requires an initial expenditure of $120,000. The project's estimated life is 8 years, and no salvage value is anticipated. The property is to be fully depreciated over the 8-year period on a straight-line basis.

Income resulting from the investment is estimated at $20,000 for the first year, $30,000 for the second year, $40,000 for the third, and $50,000 annually for the remaining 5 years. Cash flow, the total amount of money generated by the investment, is found by adding the annual depreciation charge to the after-tax profit.

The payout period answers the single question of how soon the $120,000 investment will be returned from the cash flow; it does not distinguish between profit and depreciation, nor does it consider what happens after the initial investment has been recovered.

The cumulative cash flow by years for this project is as summarized in Table 3. The results are plotted graphically in Figure 7.

Table 3 also shows the discounted value (or present worth) of the cash flows, assuming a 10-percent discount factor. When the interest is ignored, the project pays out some time between the fourth and fifth years. By inter-

Table 2 Financial Summary of Proposed Project

Year	Cash Outlay	Net Income before Taxes and Depreciation	Depreciation Charges	Net Taxable Income	Income Taxes	Net Income after Taxes	Net Cash Flow
0	(120,000)	0	0	0	0	0	(120,000)
1	0	20,000	15,000	5,000	2,500	2,500	17,500
2	0	30,000	15,000	15,000	7,500	7,500	22,500
3	0	40,000	15,000	25,000	12,500	12,500	27,500
4	0	50,000	15,000	35,000	17,500	17,500	32,500
5	0	50,000	15,000	35,000	17,500	17,500	32,500
6	0	50,000	15,000	35,000	17,500	17,500	32,500
7	0	50,000	15,000	35,000	17,500	17,500	32,500
8	0	50,000	15,000	35,000	17,500	17,500	32,500
	(120,000)	340,000	120,000	220,000	110,000	110,000	230,000
							−120,000
							110,000

Table 3 Cash Flow Summary (Project A)

Year	Net Cash Flow	Cumulative Cash Flow	Net Cash Flow Discounted at 10%	Cumulative DCF
0	(120,000)	(120,000)	(120,000)	(120,000)
1	17,500	(102,500)	15,900	(104,100)
2	22,500	(80,000)	18,600	(85,500)
3	27,500	(52,500)	20,600	(64,900)
4	32,500	(20,000)	22,200	(42,700)
5	32,500	12,500	20,200	(22,500)
6	32,500	45,000	18,300	(4,200)
7	32,500	77,500	16,700	12,500
8	32,500	110,000	15,200	27,700

polation the payoff period can be estimated at 4.6 years (or determined graphically) in this situation.

But when the time value of money is taken into account, the payout time is longer. With a 10-percent discount rate, the payoff period is extended to about 6.25 years.

Occasionally, the reciprocal of the payout time is used as another measure of investment desirability, representing the average annual cash recovery during the payout period. The 4.6-year payout indicates an average annual cash return of 1/4.6, or 21.7 percent of the initial investment. Similarly, the 6.25-year payout period yields an average cash return of 16.0 percent annually.

A major disadvantage of the payout period as a meaningful index of investment desirability is evident when another investment proposal is considered, similar in all respects to the previous example but having a longer life.

This new project (project B, summarized in Table 4) requires the same cash outlay as project A and offers the same net income before taxes and depreciation over its first 8 years. Project B, though, is expected to have an economic life of 12 years, so it will produce a \$50,000 annual income for an additional 4 years beyond the point where project A terminates. Depreciation,

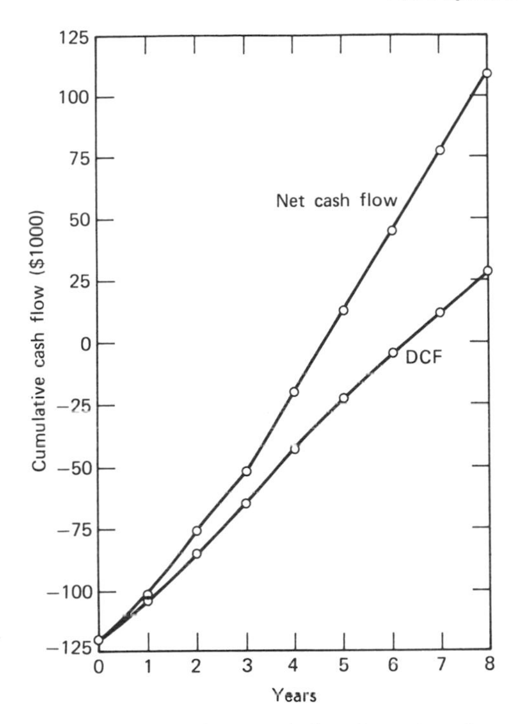

Fig. 7. *Cumulative cash flow for project A.*

still computed on the same straight-line basis, will be spread over 12 years instead of 8, thus increasing the net taxable income and decreasing the net cash flow each year.

Project B, over its 12-year life, will generate nearly twice the net cash flow of project A with its 8-year life—$210,000, compared to just $110,000 for the first project. In looking at project B's cash flow summary in Table 5 (shown graphically in Figure 8), however, it is apparent that the payout period will be slightly longer than for project A. By figuring project B's payout times on the same bases used for project A, project B is found to pay out in 5.0 years on a net cash flow basis, or in 7.0 years on a 10-percent discounted basis. Taking the reciprocals of these payout times gives average annual cash recoveries of 20.0 percent on a net basis, and 14.3 percent on a discounted basis.

In summary, the payout period is often used and referred to as a "quick-and-dirty" means of looking at the relative attractiveness of investment proposals. But by itself the payout time is *so* dirty that the analyst cannot see what lies underneath—or worse still, sees the wrong thing. Therefore the payout period should never be used as the sole criterion in investment evalua-

Table 4 Financial Summary of Project B

Year	Cash Outlay	Net Income before Taxes and Depreciation	Depreciation Charges	Net Taxable Income	Income Taxes	Net Income after Taxes	Net Cash Flow
0	(120,000)	0	0	0	0	0	(120,000)
1	0	20,000	10,000	10,000	5,000	5,000	15,000
2	0	30,000	10,000	20,000	10,000	10,000	20,000
3	0	40,000	10,000	30,000	15,000	15,000	25,000
4	0	50,000	10,000	40,000	20,000	20,000	30,000
5	0	50,000	10,000	40,000	20,000	20,000	30,000
6	0	50,000	10,000	40,000	20,000	20,000	30,000
7	0	50,000	10,000	40,000	20,000	20,000	30,000
8	0	50,000	10,000	40,000	20,000	20,000	30,000
9	0	50,000	10,000	40,000	20,000	20,000	30,000
10	0	50,000	10,000	40,000	20,000	20,000	30,000
11	0	50,000	10,000	40,000	20,000	20,000	30,000
12	0	50,000	10,000	40,000	20,000	20,000	30,000
	(120,000)	540,000	120,000	420,000	210,000	210,000	330,000
							−120,000
							210,000

Table 5 Cash Flow Summary (Project B)

Year	Net Cash Flow	Cumulative Cash Flow	Net Cash Flow Discounted at 10%	Cumulative DCF
0	(120,000)	(120,000)	(120,000)	(120,000)
1	15,000	(105,000)	13,600	(106,400)
2	20,000	(85,000)	16,500	(89,900)
3	25,000	(60,000)	18,800	(71,100)
4	30,000	(30,000)	20,500	(50,600)
5	30,000	0	18,700	(31,900)
6	30,000	30,000	16,900	(15,000)
7	30,000	60,000	15,400	400
8	30,000	90,000	14,000	14,400
9	30,000	120,000	12,700	27,100
10	30,000	150,000	11,600	38,700
11	30,000	180,000	10,500	49,200
12	30,000	210,000	9,600	58,800

tion and can seldom be justified for any use other than as an interesting, even if irrelevant, piece of information.

THE RATE OF RETURN CONCEPT

The rate of return approaches to investment evaluation measure the economic worth of an investment by relating the project's anticipated earnings to the amount of capital tied up during the project's estimated life. The main disadvantage of the rate of return approach is its relative complexity, but this disadvantage is usually outweighed by the increased precision, thoroughness, and objectivity afforded by the method.

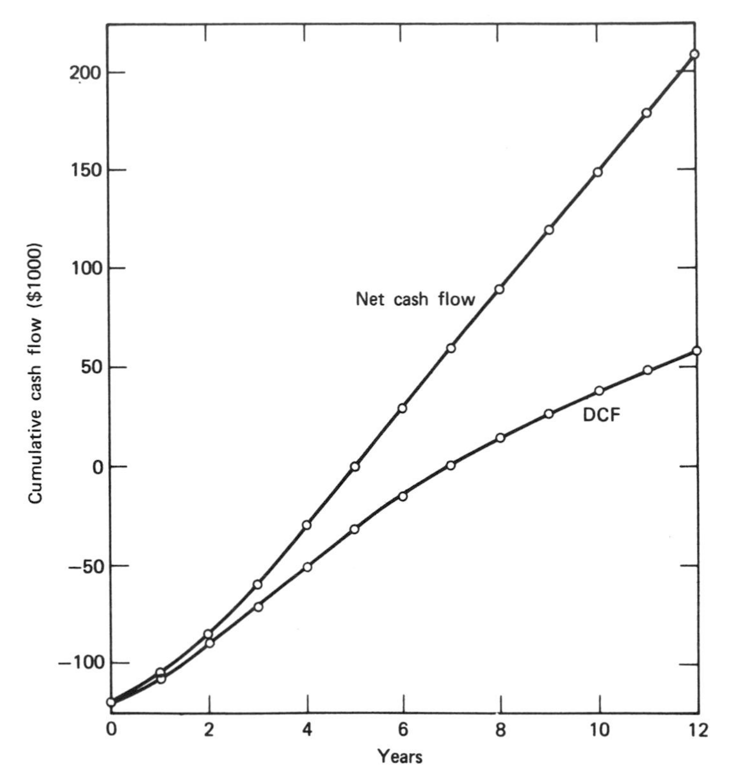

Fig. 8. *Cumulative cash flow for project B.*

A major difficulty in employing the rate of return technique is the variety of different methods in calculating the return. There are at least four approaches commonly used in determining the rate of return of a capital investment in a project, each having variations of its own. The four most widely employed methods are:

1. The accounting approach.
2. The operating return approach.
3. The present worth approach.
4. The discounted cash flow (DCF) approach.

THE ACCOUNTING APPROACH. In the conventional accounting approach to rate of return calculations, the net profit after depreciation and taxes is related to the total or average outstanding investment. This approach assumes that capital recovered through depreciation charges becomes available for use in other projects and should no longer be charged to or against the original project.

In referring again to project A, described in Table 2, the net profit after taxes over the project's 8-year economic life totals $110,000, an average of $13,750 per year. Relating this average net profit to the average outstanding investment of $60,000 results in a return on average investment of 13,750/60,000, or 22.9 percent annually. If, as is sometimes done, the earnings were related to the total initial capital outlay instead of to the average investment, the indicated return would be cut in half, to 11.5 percent.

Similarly, project B (summarized in Table 4) is found to net an average after-tax profit of $210,000/12, or $17,500 annually. This represents an annual return on average investment of 29.2 percent, or a return on total investment of 14.6 percent.

The principal shortcoming of these accounting procedures is that the time value of money is not considered. This drawback may result either in passing up desirable projects or in unknowingly entering into ventures offering low returns. Any time the income pattern, capital outlay schedule, or project life is likely to vary, the rate of return should somehow recognize the timing, as well as the magnitude, of cash expenditures and receipts.

THE OPERATING RETURN APPROACH. This method expresses the rate of return as the ratio of the average annual cash return to the original investment. The operating return, then, is a measure similar in principle to the accounting method, and as such is subject to many of the same limitations. The operating return approach is used primarily in measuring the relative operating effectiveness of different parts of a business in terms of each operation's gross cash contributions to the business as a whole. But the measure should be recognized as one of operating efficiency, *not* of economic value.

The operating return can be measured in several different ways, relating either gross operating profits (before depreciation and taxes) or cash flow to either the initial or average investment.

For the project described in Table 2 (project A), the average cash inflow over the 8-year period is $28,750 per year ($230,000/8). The operating rate of return on the original $120,000 investment, then, is 23.9 percent; and the return on the average $60,000 investment is 47.9 percent. Using the average operating profit of $42,500 ($340,000/8) instead of the cash flow gives comparable returns of 35.4 percent and 70.8 percent, depending on whether the total or average investment is used as a base.

For project B (Table 4) the average cash inflow of $330,000/12, or $27,500, results in an apparent return of 22.9 percent on the initial investment or 45.8 percent on the average investment. Similarly, the average gross operating profit ($45,000) indicates returns of 37.5 percent on the original investment and 75.0 percent on the average investment.

As with all other approaches to investment evaluation described thus far, the operating return overlooks the timing of future cash flows. For this reason it should never be used when a true measure of profitability is needed.

THE PRESENT WORTH METHOD. The present worth approach focuses on the overall cash consequences of an investment, considering both the amount and timing of cash inflows and outflows. The underlying principle of all present value methods is that money in hand is worth more than money to be received sometime in the future. The objective of present value analysis is simply to determine the value today of future cash flows generated by the investment opportunity over its economic life. This can be done by selecting an appropriate "cost" or "value" of money and using the corresponding discount factors to reduce future amounts to their present worth (as explained in the preceding chapter).

If 10 percent represents the value or cost of money to an investor, for example, $32,500 to be received in 4 years is worth only $22,200 now, as indicated in Table 6. In the same table it can be seen that the total present value of all the cash inflows for project A, discounted at 10 percent, is $147,700 or $27,700 more than the investment of $120,000 made at the beginning of the project. If the company's minimum acceptable return in this situation were 10 percent, then, it would appear desirable to commit $120,000 to this project. Or, stated another way, with money worth 10 percent, the investor should be willing to pay $147,700 for the opportunity to generate these cash returns; and since only $120,000 is required, it represents a profitable opportunity.

In Table 6 the net cash flows for project A are discounted at several different rates. In this situation, in which cash outflows occur early in the project, raising the discount rate decreases the present worth of the project. Project A's discounted net worth is depicted graphically in Figure 9, beginning with a net worth of $110,000 on a net (or nondiscounted) basis and dropping to −$17,500 to a 20-percent discount rate.

As an alternative to discounting the project's cash flows, the net profit after taxes is sometimes discounted. This is shown in Table 7, and the relationship between the present worth of discounted net profits and the discount rate is illustrated in Figure 9.

Taking the ratio of discounted earnings to the initial investment results in a *benefit/cost ratio*, a term frequently used in connection with some government projects. For tax-free public projects in which this approach is most often employed, using the discounted cash flows gives essentially the same ratio.

Table 6 Present Worth of Net Cash Flow at Various Discount Rates (Project A)

Year	Net Cash Flow	Present Worth of Net Cash Flow			
		5% Rate	10% Rate	15% Rate	20% Rate
0	(120,000)	(120,000)	(120,000)	(120,000)	(120,000)
1	17,500	16,600	15,900	15,200	14,600
2	22,500	20,400	18,600	17,000	15,600
3	27,500	23,800	20,600	18,100	15,900
4	32,500	26,700	22,200	18,600	15,700
5	32,500	25,400	20,200	16,200	13,100
6	32,500	24,200	18,300	14,100	10,900
7	32,500	23,100	16,700	12,200	9,100
8	32,500	22,000	15,200	10,600	7,600
Total in	230,000	182,200	147,700	122,000	102,500
Total out	120,000	120,000	120,000	120,000	120,000
Net	110,000	62,200	27,700	2,000	(17,500)

The same information given for project A is summarized for project B in Tables 8 and 9, and in Figure 10. While the same general relationships hold, project B's longer economic life results in a higher present worth than for project A at any given discount rate. Project B's present worth ranges from \$210,000 on a nondiscounted basis to −\$8800 at a 20-percent rate. The present worth of its net profits go from \$210,000 down to \$66,400 over the same range of interest rates.

Even though calculations for the present worth method are straightforward and result in a valid measure of investment worth, there are still two main problems associated with this approach:

1. Agreement on what constitutes an appropriate discount rate or minimum acceptable return on investment is sometimes difficult.

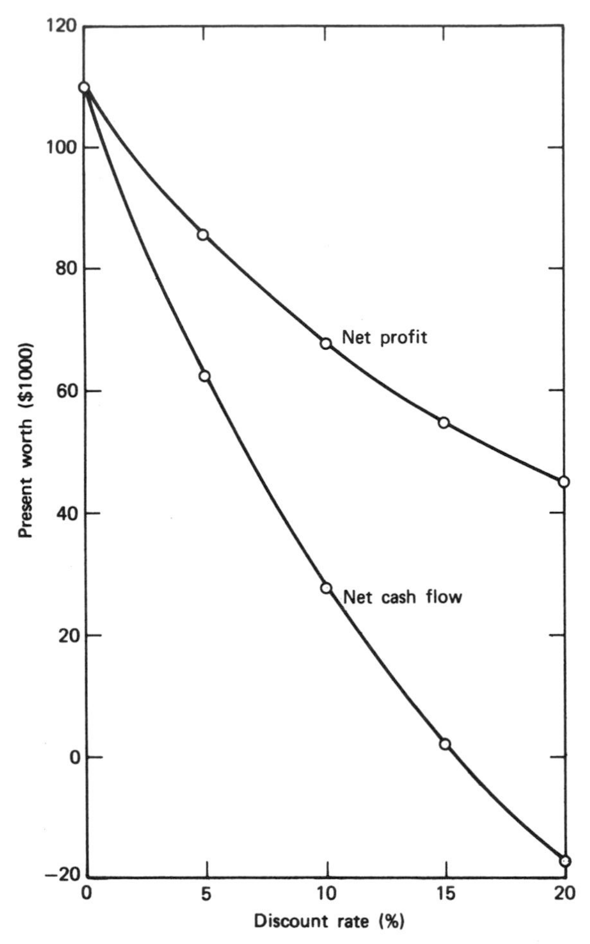

Fig. 9. *Present value of future profits and cash flows for project A.*

2. The answer is expressed in dollars, rather than as a percentage return on invested capital; thus comparisons among projects having different investment requirements may be awkward.

THE DCF APPROACH. The DCF approach is a special case of the present worth method, in which the sum of the present worths of all cash flows—both in and out—is set equal to zero. Whatever discount rate applied to the cash flows makes their discounted values total zero is defined as the DCF rate of return. As might be expected, solving for the appropriate discount

Table 7 **Present Worth of Net Profit after Taxes at Various Discount Rates (Project A)**

Year	Net Profit after Taxes	Present Worth of Net Profit			
		5% Rate	10% Rate	15% Rate	20% Rate
1	2,500	2,400	2,300	2,200	2,100
2	7,500	6,800	6,200	5,700	5,200
3	12,500	10,800	9,400	8,200	7,200
4	17,500	14,400	12,000	10,000	8,400
5	17,500	13,700	10,900	8,700	7,100
6	17,500	13,100	9,950	7,600	5,900
7	17,500	12,400	9,000	6,600	4,900
8	17,500	11,900	8,200	5,700	4,100
Total	110,000	85,500	67,950	54,700	44,900

rate is mathematically complex (if not impossible), so most analysts employ either trial-and-error or graphical techniques in the solution of DCF problems. Most computer programs work the same way but much faster.

The rate of return calculated by using the DCF technique represents the rate of return on invested capital with continuous compounding; it assumes that earnings generated by the capital investment are reinvested in the project to continue earning at the same rate. Similar to the other present worth methods, the DCF approach is concerned with both the magnitude and timing of all cash outlays and receipts. Additionally, the DCF method avoids the two main problems inherent in the other present worth methods by eliminating the need for selecting a sometimes-arbitrary interest rate and by expressing the results as a percentage return on investment.

The mechanics of the DCF approach can best be visualized by referring to Table 6 in which project A's cash flows are shown discounted at different rates. In Table 6 it can be seen that the present worth of the project's net cash flows is $2000 at a 15-percent discount rate and drops to −$17,500 at a 20-percent rate. Since the DCF return is defined as the discount rate at which the project's net present worth is zero, the answer obviously lies between 15 percent

Table 8 Present Worth of Net Cash Flow at Various Discount Rates (Project B)

Year	Net Cash Flow	Present Worth of Net Cash Flow			
		5% Rate	10% Rate	15% Rate	20% Rate
0	(120,000)	(120,000)	(120,000)	(120,000)	(120,000)
1	15,000	14,300	13,600	13,000	12,500
2	20,000	18,100	16,500	15,100	13,900
3	25,000	21,600	18,800	16,500	14,500
4	30,000	24,700	20,500	17,200	14,500
5	30,000	23,500	18,700	14,900	12,100
6	30,000	22,400	16,900	13,000	10,100
7	30,000	21,300	15,400	11,300	8,400
8	30,000	20,300	14,000	9,800	7,000
9	30,000	19,400	12,700	8,500	5,800
10	30,000	18,400	11,600	7,400	4,900
11	30,000	17,600	10,500	6,500	4,100
12	30,000	16,700	9,600	5,600	3,400
Total in	330,000	238,300	178,800	138,800	111,200
Total out	−120,000	−120,000	−120,000	−120,000	−120,000
Net	210,000	118,300	58,800	18,800	(8,800)

Table 9 Present Worth of Net Profit after Taxes at Various Discount Rates (Project B)

Year	Net Profit after Taxes	Present Worth of Net Profit			
		5% Rate	10% Rate	15% Rate	20% Rate
1	5,000	4,800	4,600	4,400	4,200
2	10,000	9,100	8,300	7,600	6,900
3	15,000	13,000	11,300	9,900	8,700
4	20,000	16,500	13,700	11,400	9,600
5	20,000	15,700	12,400	9,900	8,000
6	20,000	14,900	11,300	8,600	6,700
7	20,000	14,200	10,300	7,500	5,600
8	20,000	13,500	9,300	6,500	4,700
9	20,000	12,900	8,500	5,700	3,900
10	20,000	12,300	7,700	4,900	3,200
11	20,000	11,700	7,000	4,300	2,700
12	20,000	11,100	6,400	3,700	2,200
Total	210,000	149,700	110,800	84,400	66,400

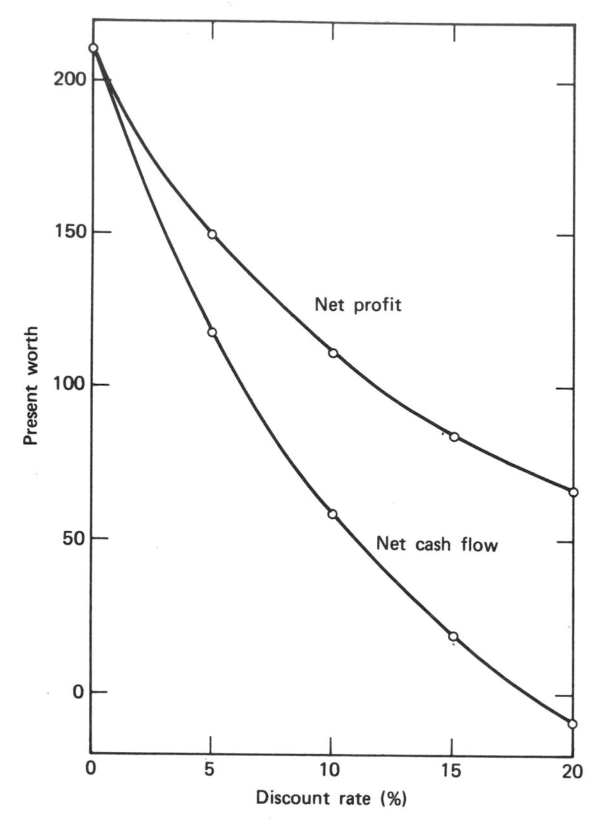

Fig. 10. *Present worth of future profits and cash flows for project B.*

and 20 percent. Interpolation can be used to estimate the DCF return, or a graphical method can be employed as in Figure 9. In Figure 9 the point at which the present-worth-of-net-cash-flow curve crosses the zero line identifies the DCF return—approximately 15.5 percent in this case. Similarly, by referring to Table 8 and Figure 10, the DCF return on project B is found either graphically or by interpolation to be 18.0 percent.

The return on investment found in this manner may be known as the internal rate of return, the investor's rate of return, the profitability index, or the DCF return. Whatever it is called, if the indicated rate offers an acceptable return to the investor, the investment proposal presumably will be approved. If the indicated return falls below the investor's minimum acceptable return, then the proposal will be rejected. If several alternative investments are available, those showing the highest percentage returns will be given top priority.

Table 10 Comparison of Different Investment Evaluation Techniques

Method of Evaluation	Project A (Table 2)	Project B (Table 4)
Payout time		
Net	4.6 years	5.0 years
Discounted at 10%	6.3 years	7.0 years
Average annual payout		
Net	21.7%	20.0%
Discounted at 10%	16.0%	14.3%
Accounting return		
Average net profit on original investment	11.5%	14.6%
Average net profit on average investment	22.9%	29.2%
Operating return		
Average operating profit on original investment	35.4%	37.5%
Average operating profit on average investment	70.8%	75.0%
Average cash flow on original investment	23.9%	22.9%
Average cash flow on average investment	47.9%	45.8%
Present worth of net cash flow		
Discounted at 10%	$27,700	$58,800
Discounted at 20%	($17,500)	($ 8,800)
Present worth of net profit		
Discounted at 10%	$67,900	$110,800
Discounted at 20%	$44,900	$66,400
Discounted cash flow return	15.5%	18.0%

The main objections to the DCF approach are:

1. The inference that cash generated by the investment can be reinvested in the project, or at least will continue to earn at the same rate calculated for the project under consideration.

2. The complexity of the calculations involved in computing the rate of return on invested capital.

The first objection is not necessarily valid, however, since the primary purpose of the DCF analysis is simply to reflect the relative productiveness of the capital being committed to the project under consideration, *not* to provide an absolute measure of profitability. The resulting index of capital productivity does provide a valid, objective basis for comparing alternative uses of capital.

The second objection—complexity—can be overcome by the use of the many available computer programs and short-cut methods designed for solving DCF problems.

SUMMARY

The attractiveness of a proposed investment depends primarily upon the value the investor places on his money and the manner in which he defines that value. Table 10 summarizes the results of the different methods of investment evaluation described in this chapter as applied to two different projects. Either project may appear preferable, depending on which evaluation technique is employed. The wide range of values shown in Table 10 merely emphasizes the necessity for defining the approach to be used and the criteria to be employed prior to conducting an investment analysis. Of this multitude of approaches, the present value and the DCF techniques offer the only valid measures of investment profitability since they, unlike other methods, consider both the amount and timing of all cash inflows and outflows. The discounted cash flow method is usually preferable, since it expresses profitability in a single percentage figure which is applicable regardless of the investor's cost of money. Any approach that fails to consider the time value of money cannot be considered a true measure of investment worth.

4

Cash Flow Analysis

The preceding chapter discussed the various ways in which an investment proposal might be evaluated. It was concluded that, of the many ways possible, only approaches based on the present value of future cash flows afford a realistic measure of investment desirability.

This chapter deals with the analysis of cash flows, employing the DCF technique in calculating return on investment (ROI). The ROI thus determined serves as a useful and objective measure of the merits of an individual investment or of the relative desirability of several alternative investments. ROI, as an analytical measure of investment worth, provides management with a sound basis for approval (or disapproval) of an investment proposal, or for choosing the best of several opportunities.

THE DCF CONCEPT

Making a profitable investment is not especially difficult once the investment opportunity has been pointed out; but identifying a profitable investment opportunity, or the best of several opportunities, requires a considerable amount of mathematical prophecy.

Most engineering projects involve a choice between alternatives having different investment requirements and perhaps covering different time periods. So that the best possible alternative can be selected, it is important to have an objective means of evaluating all alternatives on a consistent and comparable basis.

The DCF approach is generally recognized as a logical, and by many as the *most* logical, way to analyze and compare alternative investment opportunities involving different time periods and generating different cash (or benefit) flows.

The term cash flow, as used here, is defined as the net profit after taxes, plus depreciation, depletion, and other "paper" expenses; thus cash flow represents the total amount of money generated by the project and available for other uses. Cash outflows are actual out-of-pocket expenditures; cash inflows are similar to money in the bank. The net cash flow, then, is simply the difference between the cash inflows and outflows. Cash flow is essentially a function of price, cost, volume, investment requirement, depreciation, and tax structure.

It must be emphasized that cash flow is *not* the same as profit. Cash flow represents only the difference between total cash receipts from the business's operations and the total cash outlays incurred in carrying out these operations. Depreciation, which must be considered in calculating profit, is not a cash expense and therefore affects cash flow only by reducing the out-of-pocket income tax liability.

In DCF analysis a project's net cash flow is estimated for each year of its projected economic life. Then these cash flows are discounted at an interest rate calculated to make the sum of discounted cash inflows equal to the sum of discounted cash outflows.

The interest rate that results in the discounted cash inflows being equal to the discounted cash outflows represents the project's ROI, also referred to as its profitability index (PI) or the DCF rate of return. Whatever its name, this interest rate represents the highest rate at which the required capital can be borrowed and paid off from the cash earnings generated by the project over its economic life. If money were available at a lower interest rate, the debt could be retired with something left over; if the company's cost of capital were higher than the indicated rate, then the cash generated by the project would not be enough to pay off the borrowed money.

The following example illustrates the effect of the timing of cash flows on a project's ROI as measured by the DCF approach. Table 11 shows three simple 5-year projects, each requiring an initial cash outflow of $1000 and each followed by 5 years of cash inflows totaling $1500. The first case offers a cash return of $300 annually; the second has a declining pattern of cash flows, starting at $500 and dropping by $100 each year; and the third has a

Table 11 Effect of Cash Flow Timing on Project ROI

Year	Net Cash Flow Project No. 1	Project No. 2	Project No. 3
0	−$1000	−$1000	−$1000
1	300	500	100
2	300	400	200
3	300	300	300
4	300	200	400
5	300	100	500
ROI	15%	20%	12%

first-year cash return of $100, increasing by $100 each year to $500 in the fifth year.

While the net totals of cash inflows and outflows are the same in each of the three cases—$1000 out and $1500 in—there is considerable difference in the relative desirability of these projects. Obviously, project 2 is preferred since it returns the investment most quickly, while project 3 appears least desirable for the opposite reason. As measured by the DCF technique, project 1 has a 15-percent ROI; project 2, a 20-percent ROI; and project 3 yields a 12-percent return. Thus the DCF approach objectively puts the projects in the same order as is intuitively obvious. This is perhaps one of the fundamental benefits of DCF analysis—especially when the results are less obvious.

PROBLEMS IN DCF ANALYSIS

A chief drawback in using the DCF method of analysis is the difficulty involved in finding the appropriate interest rate; unless a computer program is available, many tedious trial-and-error calculations may be necessary. For example, consider the cash flow schedules described in Table 11.

The conventional trial-and-error approach to solving these problems is to first total the net cash inflows and outflows. Since total inflows are greater than total outflows, this indicates that the projects are earning some return on the initial investment.

Next, some interest rate (say 10 percent) is selected; each annual net cash flow is discounted by a factor appropriate to that rate, and the discounted inflows compared with the discounted outflows. At a 10-percent rate discounted inflows are still higher than discounted outflows in each case, indicating that the projects are all earning at a rate higher than 10 percent.

The same process is then repeated, using perhaps a 20-percent rate for this trial. Again discounted cash inflows are compared with discounted cash outflows. At the 20-percent discount rate, the cash flows for project 2 are found to total zero, indicating a 20-percent ROI for this project. For the other two projects, the discounted cash outflows exceed the inflows, indicating ROIs between 10 and 20 percent.

For projects 1 and 3 the next step could employ interpolation between 10 and 20 percent to approximate their true yields; or other rates between 10 and 20 percent could be tried until the correct rates are found.

In any event the arithmetic involved in DCF analysis can be tedious and is undoubtedly one of its chief drawbacks.

DOUBLE AND IMAGINARY SOLUTIONS

Because of the high-order equations involved in complex DCF analyses, it is sometimes possible for a single problem to have several solutions. The misleading results that are possible from a DCF problem involving multiple solutions are illustrated in the following situation.

Year	Net Cash Flow ($)
0	+1000
1	−3000
2	+2000

On visual examination the merits of such an investment appear highly questionable; with the total cash inflow equaling the total cash outflow, a zero-percent ROI obviously applies. This problem can be solved mathematically for the ROI by setting the sum of the DCFs equal to zero and solving for the discount rate R:

$$1000 - \frac{3000}{(1 + R)} + \frac{2000}{(1 + R)^2} = 0$$

Solving for R, the ROI:

$$\begin{aligned} 1000(1 + R)^2 - 3000(1 + R) + 2000 &= 0 \\ R^2 - R &= 0 \\ R(R - 1) &= 0 \\ R = 0;\ R &= 1 \end{aligned}$$

So, the ROI in this case can be either zero or 100 percent, both of which satisfy the DCF definition of ROI:

Year	Net Cash Flow ($)	Net Cash Flow Discounted at 0%	Net Cash Flow Discounted at 100%
0	+1000	+1000	+1000
1	−3000	−3000	−1500
2	+2000	+2000	+500
		0	0

In this case, at least, the 100-percent ROI figure can easily be seen to be suspect upon examination of the problem. While mathematically correct, it is obviously unrealistic from an investment standpoint.

But in some problems the presence, let alone the practicality, of a double solution may be far less apparent. Any trial-and-error solution, whether obtained manually or on a computer, may find one mathematically correct answer but ignore the other.

Fortunately, the possibility of such situations can be easily identified by examining the cumulative cash flow pattern. A different solution can occur each time the cumulative cash flow changes from positive to negative or from negative to positive. This does not happen in most projects since there is generally a substantial cash outflow during the initial time period, followed by a series of cash inflows over the remainder of the project's economic life. Should heavy expenditures be required toward the middle or end of the project's life, though, the possibility of a double solution must be recognized. Secondary recovery of oil is a good example, in which substantial expenditures may be required to restimulate a declining oil flow—essentially a "double investment" situation.

Imaginary solutions to DCF problems may also occur in the same types of situations that lead to double solutions. A net cash flow of −1000, +2000, −2000, for example, yields an imaginary result. Such results are unlikely to cause problems, though.

MATHEMATICAL SOLUTION OF DCF PROBLEMS

The complexity of most DCF problems precludes any direct mathematical solution. Only when a 2- or 3-year economic life is assumed, or when there is a uniform series of cash flows, can a solvable equation be formulated. The general form of a DCF problem can be expressed as:

$$NCF_0 + \frac{NCF_1}{(1 + R)} + \frac{NCF_2}{(1 + R)^2} + \frac{NCF_3}{(1 + R)^3} + \ldots + \frac{NCF_n}{(1 + R)^n} = 0$$

where NCF_n represents the net cash flow occurring in the nth year. To solve for the ROI requires that this equation be solved for R—a task far beyond the abilities of most mathematicians.

The simplifying assumption of a uniform cash flow pattern may be realistic enough for some projects, especially if the object of the analysis is simply to screen out undesirable projects or to obtain a rough approximation of a project's ROI when insufficient data are available for a more thorough analysis. In such a situation a quick mathematical solution can be quite useful.

When a single cash outflow (representing the initial capital investment) is made at the beginning of a project, followed by uniform cash inflows over the remainder of the project's economic life, the resulting ROI can be found by using the formula for computing the capital recovery factor:

$$\text{capital recovery factor} = \frac{R(1 + R)^n}{(1 + R)^n - 1} = \frac{\text{annual cash inflow}}{\text{initial investment}}$$

where the initial investment is made in year 0 and the annual cash inflow takes place uniformly over the years 1 through n.

Given the initial cash outflow, the annual cash inflows, and the economic project life, solving the above equation for R gives the ROI.

While the equation is solvable, a good set of interest tables make the job much easier. The ratio of the annual cash inflow to the initial investment gives the capital recovery factor. All that need be done is to find in the interest tables the interest rate having that capital recovery factor for the desired number of years. For example, if the capital recovery factor were 0.300 for a 6-year period, examination of interest tables would show a 20-percent interest rate to have a capital recovery factor of 0.3007057 for a 6-year uniform annual cash recovery. The ROI in this case therefore would be about 20 percent.

This approach, as employed in screening new projects, is discussed in some detail in a later section in which a graphical solution and a mathematical approximation are also presented.

TRIAL-AND-ERROR SOLUTIONS

A mathematical trial-and-error approach can be used to "bracket" a project's ROI. Then the approximate ROI can be calculated by interpolation. This procedure is commonly used and can give a satisfactory solution without a great deal of computation time. A maximum of about five trial solutions is generally sufficient.

A distinct advantage of this approach is that it can be set up in a standard format so that the necessary calculations can be performed in a routine fashion by clerical personnel.

Figure 11 shows a typical DCF worksheet. The net cash investment for each year of the project's life is inserted in the appropriate column of the upper table, and the net cash receipts in the lower table. Then the annual cash flows are discounted by multiplying each cash flow by its corresponding discount factor at the indicated 10-, 15-, 25-, and 40-percent trial interest rates. The arithmetic total of the net cash inflows and outflows represents discounting at a zero-percent rate. This is done for both cash outflows (investment) and cash inflows (receipts). Finally, the columns are totaled, and the ratio of total discounted receipts to total discounted investments is calculated.

Since the return on investment is by definition the interest rate that makes the discounted inflows equal to the discounted outflows, a ratio of discounted receipts to discounted investments of 1.000 identifies the interest rate being sought. A ratio of discounted inflows to discounted outflows greater than 1.0 indicates that the trial interest rate is too low; similarly, a ratio of less than 1.0 indicates that the trial rate is too high.

Having computed the ratios for the five trial interest rates—0, 10, 15, 25, and 40 percent—the return on investment percentage can be interpolated by means of the following formula:

$$\text{ROI} = a + (b - a)\left(\frac{a_r - 1.00}{a_r - b_r}\right)$$

where a is the trial interest rate at which the ratio of discounted receipts to discounted investments is larger than, but closest to, 1.0; b is the trial interest rate having a ratio smaller than, but closest to, 1.0; and a_r and b_r are their respective ratios.

For example, if the ratio of discounted receipts to discounted investments were found to be 1.15 at a 15-percent trial interest rate and 0.80 at a trial rate of 25 percent, the ROI would be approximately as follows.

$$\text{ROI} = 15 + (25 - 15)\frac{1.15 - 1.00}{1.15 - 0.80} = 19.3 \text{ percent}$$

The interest factors shown in Figure 11 are the standard end-of-year present value rates found in most interest tables, representing the $1/(1 + R)^n$ factor. Some companies prefer to use an average midyear factor instead; thus the discount factor applying to investments or receipts incurred during year n is $1/(1 + R)^{n-0.5}$. In either case the approach is the same, and the difference in results is not critical so long as the preferred method is applied consistently.

Year	Trial No. 1 0% Interest Rate	Trial No. 2 10% Interest Rate		Trial No. 3 15% Interest Rate		Trial No. 4 25% Interest Rate		Trial No. 5 40% Interest Rate	
	Actual Amount	Discount Factor	Present Value	Discount Factor	Present Value	Discount Factor	Present Value	Discount Factor	Present Value
0		1.000		1.000		1.000		1.000	
1		0.909		0.870		0.800		0.714	
2		0.826		0.756		0.640		0.510	
3		0.751		0.658		0.512		0.364	
4		0.683		0.572		0.410		0.260	
5		0.621		0.497		0.328		0.186	
6		0.564		0.432		0.262		0.133	
7		0.513		0.376		0.210		0.095	
8		0.467		0.327		0.168		0.068	
9		0.424		0.284		0.134		0.048	
10		0.386		0.247		0.107		0.035	
Total cash outflow									
0		1.000		1.000		1.000		1.000	
1		0.909		0.870		0.800		0.714	
2		0.826		0.756		0.640		0.510	
3		0.751		0.658		0.512		0.364	
4		0.683		0.572		0.410		0.260	
5		0.621		0.497		0.328		0.186	
6		0.564		0.432		0.262		0.133	
7		0.513		0.376		0.210		0.095	
8		0.467		0.327		0.168		0.068	
9		0.424		0.284		0.134		0.048	
10		0.386		0.247		0.107		0.035	
Total cash inflow									
Inflow/ outflow ratio									

ROI calculation

Fig. 11 Typical DCF worksheet

GRAPHICAL SOLUTION OF DCF PROBLEMS

The conventional graphical solution to DCF problems simply substitutes a graph for the mathematical interpolation described in the preceding section. Once the ratios of discounted cash inflows to discounted cash outflows have been calculated for the selected trial interest rates, the ratios are plotted on a graph. A smooth curve may then be drawn through the points, and the discount rate corresponding to the point on the curve having a ratio of 1.0 will be the ROI percentage.

Figure 12 shows how the graphical solution technique works. Here the following values were obtained for the five trial interest rates.

Trial Interest Rate (%)	Ratio of Discounted Cash Inflow to Discounted Cash Outflow
0	2.18
10	1.45
15	1.22
25	0.91
40	0.64

These five points are plotted as shown in Figure 12, with the ratios on the vertical axis and the interest (or discount) rates on the horizontal axis. A horizontal line has been drawn at a ratio of 1.0, and the point at which the curve intersects this line identifies the project's ROI. In the example the ROI is found to be 21.5 percent.

It is interesting to note that a linear interpolation between the 15- and 25-percent rates, using the equation from the preceding section, gives an answer to this same problem of 22.1 percent. Since the relationship between discounted cash flow values and interest rates is not linear, however, the graphical method is more accurate than the arithmetic approximation.

THE INVESTMENT PROFIT PROPHET

The *investment profit prophet* (Figure 13) is a graphical device for solving DCF problems.

Even complex problems involving irregular cash flow patterns as in the example shown in Table 12 can be easily solved without trial-and-error calculations. The procedure is as follows.

1. Using the 40-percent discount factors shown in Table 13, discount the net cash flows by years as illustrated in Table 14.

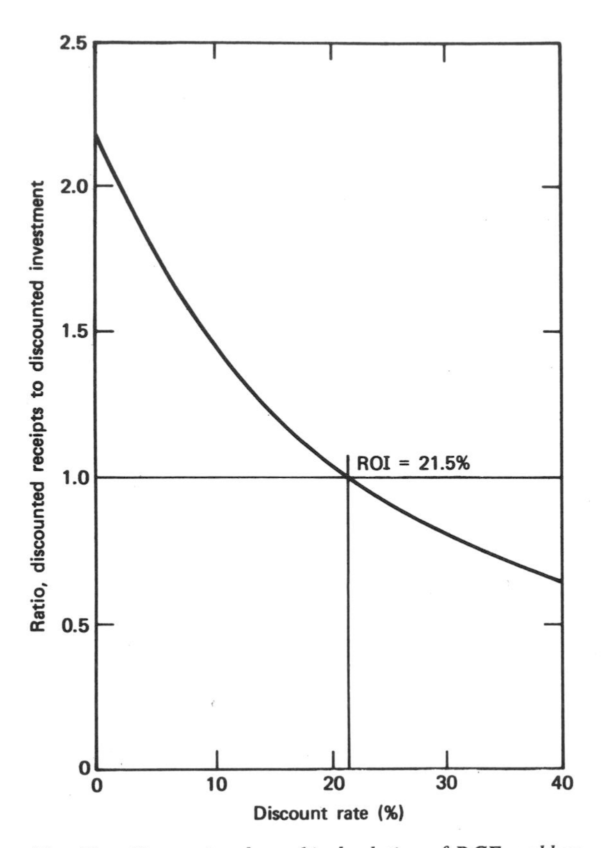

Fig. 12. *Conventional graphical solution of DCF problem.*

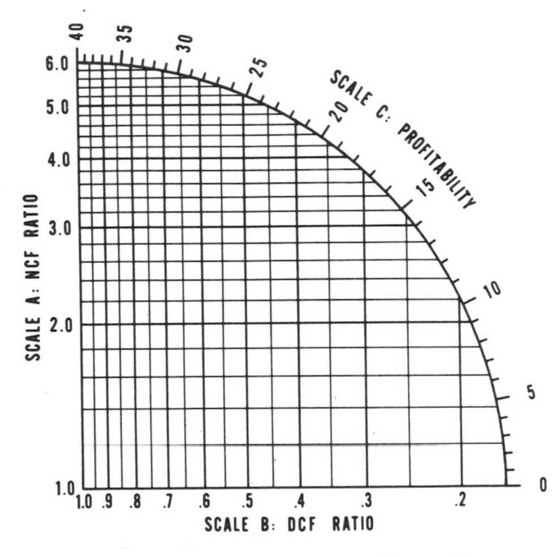

Fig. 13. *The investment profit prophet.*

Table 12 Cash Flow Summary for Profit Prophet Example

Year of Operation	Net Profit after Taxes	Depreciation	Investment Requirement	Net Cash Flow
0	(50,000)	0	550,000	(600,000)
1	(175,000)	275,000	1,300,000	(1,200,000)
2	20,000	230,000	150,000	100,000
3	160,000	190,000	150,000	200,000
4	290,000	160,000	150,000	300,000
5	270,000	130,000	0	400,000
6	290,000	110,000	0	400,000
7	310,000	90,000	0	400,000
8	310,000	90,000	0	400,000
9	310,000	90,000	0	400,000
10	310,000	90,000	0	400,000
11	310,000	90,000	0	400,000
12	310,000	90,000	(600,000)	1,000,000

2. Compute the net cash flow ratio, the ratio of net cash in to net cash out; in the example the ratio is 2.44 (Table 14).

3. Compute the DCF ratio, the ratio of discounted cash in to discounted cash out, discounted at the 40-percent rate; in the example the DCF ratio is 0.293 (Table 14).

4. Locate the intersection of the net cash flow ratio (scale A) and the DCF ratio (measured on scale B) on the *profit prophet* graph. (See Figure 14.)

5. Extend a line from the origin through this point of intersection and read the return on investment on scale C. As shown in Figure 14, the ROI is found to be between 13 and 14 percent.

The *profit prophet* is applicable to projects having an economic life of from 5 to 15 years and yielding a return between 0 and 40 percent; the *profit prophet*

Table 13 Forty-Percent Discount Factors for Use with Investment Profit Prophet

Year	Amount
0	1.000
0.5	0.844
1.0	0.714
1.5	0.604
2.0	0.510
2.5	0.431
3.0	0.364
3.5	0.308
4.0	0.260
4.5	0.220
5.0	0.186
6.0	0.133
7.0	0.095
8.0	0.068
9.0	0.048
10.0	0.035
Over 10 years: use 0.000	

simply provides a quick means of graphic interpolation when the interest rate and the project life fall within these ranges. Its accuracy is adequate for most purposes, with the indicated answers falling within a percent or so of the true ROI. If a greater degree of accuracy is desired, the ROI given by the *profit prophet* can be used as a trial interest rate and the true rate can be quickly established.

Table 14 Return on Investment Computation for Profit Prophet Example*

Year of Operation	Net Cash Flow	40% Discount Factor	DCF
0	(600,000)	1.000	(600,000)
1	(1,200,000)	0.714	(857,000)
2	100,000	0.510	51,000
3	200,000	0.364	73,000
4	300,000	0.260	78,000
5	400,000	0.186	74,000
6	400,000	0.133	53,000
7	400,000	0.095	38,000
8	400,000	0.068	27,000
9	400,000	0.048	19,000
10	400,000	0.035	14,000
11	400,000	0.000	0 0
12	1,000,000	0.000	0

*Net cash flow ratio (net cash in/net cash out) = 4,400,000/1,800,800 = 2.44. DCF ratio (discounted cash in/discounted cash out) = 427,000/1,457,000 = 0.293. ROI (see Figure 14) = approximately 14%.

SCREENING NEW PROJECTS

Complicated problems may require complicated solutions, and complex questions may require complex answers. But complexity should be avoided whenever possible. Complex solutions may guarantee complex answers, but they cannot result in a level of precision beyond that of the data employed.

Most companies are interested in pursuing whatever profitable new opportunities may come along. Frequently, there are many different opportunties under consideration at the same time. While most are probably not worth

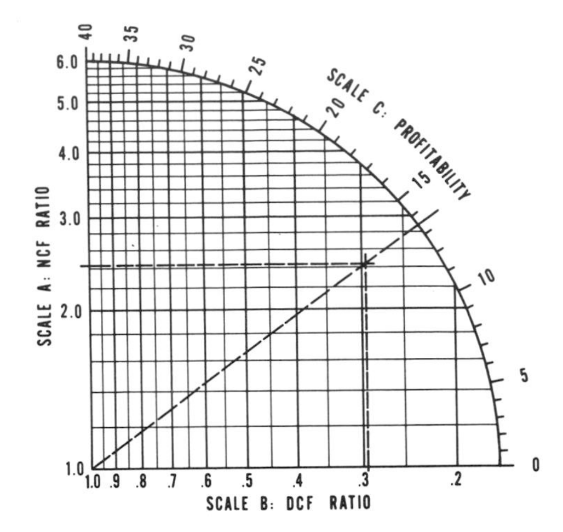

Fig. 14. *Example of ROI determination using profit prophet graph.*

following, a few may be profitable and an occasional one may be especially rewarding. How to classify all the available opportunities and to distinguish from among them the most promising is an often perplexing management problem.

For purposes of classification it is especially important to keep economic studies simple in their early stages. The accuracy of available economic data may preclude any precise evaluation and seldom justifies even a serious attempt at precision.

As early as possible in an economic investigation, though, simple calculations should be carried out to determine whether or not the overall concept being considered is worthy of further study. The data can be refined and the details can be filled in later. But it is seldom desirable to undertake any elaborate or sophisticated analysis in a project's early stages with the thought of obtaining a complete answer on all points.

The "screening test" is a type of analysis aimed simply at determining whether a particular proposal is worth pursuing further, or at finding which of several alternatives may offer the best opportunities for further investigation.

A "Programmed" Approach to Project Screening

Table 15 offers a programmed approach to screening new projects (the program is written for human beings rather than computers). It requires only six input figures. It results in an approximate ROI percentage, employing the DCF approach.

Table 15 Project Evaluation Program

Line	Item	Units	Source	Example A	Example B
1	Average annual sales	$	Input	200,000	200,000
2	Direct production costs	$	Input	100,000	120,000
3	Indirect and overhead costs	$	Input	40,000	50,000
4	Net investment	$	Input	100,000	80,000
5	Economic project life	Years	Input	5	8
6	Income tax rate	%	Input	50	50
7	Average depreciation	$	Line 4/line 5	20,000	10,000
8	Total deductions	$	Line 2 + line 3 + line 7	160,000	180,000
9	Net profit before taxes	$	Line 1 line 8	40,000	20,000
10	Income taxes	$	Line 6 x line 9	20,000	10,000
11	Net profit after taxes	$	Line 9 – line 10	20,000	10,000
12	Net cash flow	$	Line 11 + line 7	40,000	20,000
13	Capital recovery rate	Ratio	Line 12/line 4	0.400	0.250
14	ROI	%	Graph	28.5	18.6

The six inputs required in this program and the values used for each in the two examples shown in Table 15 are:

Line	Item	Example A	Example B
1	Average annual sales or revenues	$200,000	$200,000
2	Direct production costs	$100,000	$120,000
3	Indirect and overhead costs	$ 40,000	$ 50,000
4	Net investment	$100,000	$ 80,000
5	Economic project life	5 years	8 years
6	Income tax rate	50%	50%

Average annual sales (line 1) require that an estimate be made of the number of units sold at a given price, less direct selling costs and commissions.

Similarly, *direct production costs* (line 2) refer primarily to the unit costs of labor and materials incurred at the assumed production volume.

Indirect and overhead costs (line 3) are generally accrued on a time basis rather than on a per-unit-of-production basis; they include capital charges, administrative and general expenses, and other time related costs of carrying on the proposed venture.

Net investment (line 4) refers to the total capital required to put the venture in the startup stage, less any appropriate tax credits. Working capital requirements need not be considered in this initial screening.

Economic project life (line 5) refers to the number of years during which the assumed conditions will persist and over which the original investment must be recovered. A project life of between 5 and 10 years is commonly used. Less than 5 years is usually unrealistic, and conditions beyond 10 years can seldom be anticipated.

Income tax rate (line 6) refers to the total of all income-based taxes—federal, state, and local—plus any surcharges that might apply.

Lines 7 through 13 in the screening program consist of the calculations necessary to determine the capital recovery rate (line 13), which can be translated directly into a percentage ROI by reference to the graph (Figure 15) or to interest tables containing capital recovery factors for a wide range of values. The graph is entered on the vertical axis at the computed capital recovery rate, and the ROI is read on the horizontal scale for the appropriate project life.

Arithmetic Approximation of the ROI

To calculate the *approximate* return on investment without reference to either a graph or interest tables, the following three lines can be added to the program.

Line	Item	Units	Source	Example A	Example B
15	Depreciation rate	Ratio	1.00/line 5	0.200	0.125
16	Excess return rate	Ratio	Line 13 Line 15	0.200	0.125
17	Approximate return on investment	%	150 × line 16	30	19

These approximate ROI figures compare favorably with the values obtained graphically: 30 percent against 28.5 percent, and 19 percent against 18.6 percent.

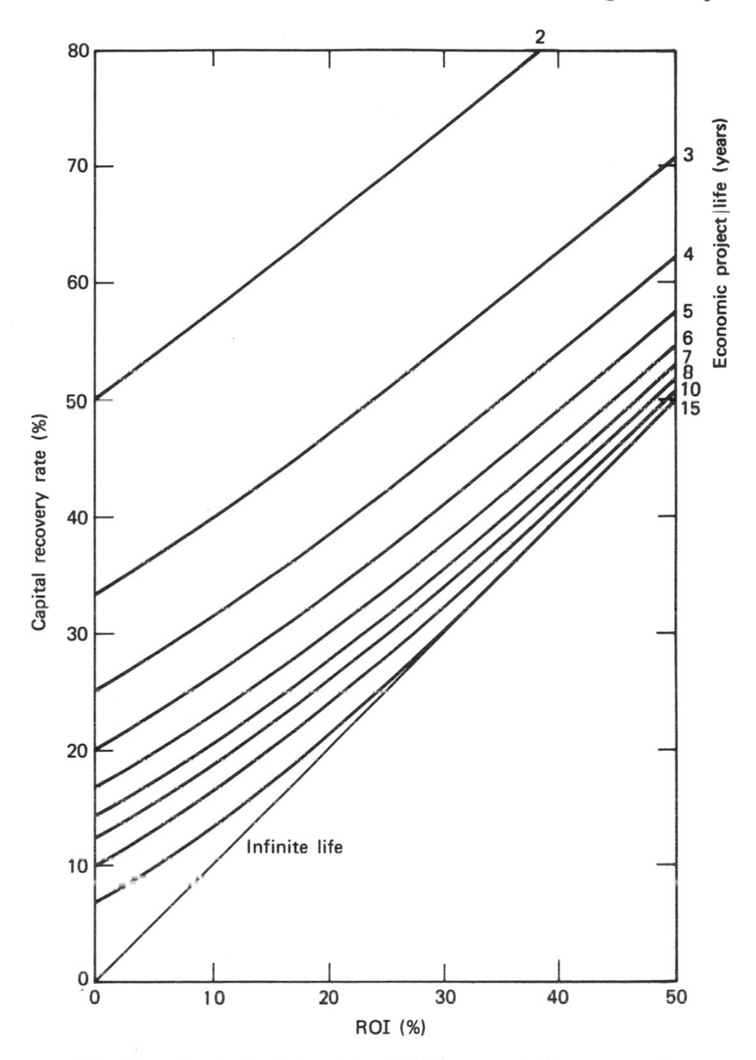

Fig. 15. *Graph for determining ROI from capital recovery rate.*

Assumptions

Two important assumptions are necessary in this highly simplified approach to project evaluation:

1. Annual sales, costs, and tax rates remain constant over the project's economic life.
2. The entire investment is depreciated (or amortized) on a uniform straight-line basis over the project's economic life.

The program presented here gives an accurate solution to the problem under the conditions imposed by these simplifying assumptions. If the assumptions are tolerable, then the answers will be adequate for their intended purpose.

This is a quick way to get a good "feel" for the economic attractiveness of a proposed project in its early stages.

SENSITIVITY ANALYSIS FOR NEW PROJECTS

Sensitivity can be roughly defined as the change in output brought about by a specified change in input. The main purpose of sensitivity analysis is to determine which of several inputs may be the most important and the extent to which each input affects the output.

The project evaluation program presented in the preceding section offers a convenient means of identifying the sensitivity of a project's ROI to changes in annual revenues, direct and indirect costs, economic life, capital investment requirements, and income tax rates. Thus the relative importance of different economic parameters can be assessed during the early planning stages of a project, before any major commitments are made. Additional investigation can then be directed toward clarifying the most relevant factors.

While the program described in Table 15 requires six inputs, variations in only three of the six inputs need be considered for most projects. Since revenues, direct costs, and indirect costs are eventually combined, they can be considered together as a single parameter; only the net difference between revenue and cost is significant. In the following example, costs are held constant while the sales revenue is varied to generate the different levels of net operating profit—the net amount of cash left over after all operating costs and expenses have been paid but before deductions for depreciation and taxes have been taken. The same effect could be obtained by holding the sales constant while varying the cost figures.

In looking at new projects, there are many uncertainties involved and it is often difficult to define all the conditions under which the project must operate. Predesign estimates of capital requirements might be off by as much as 30 percent, while estimates of economic life are always uncertain since they depend upon a number of uncontrollable external, as well as internal, factors. Income tax considerations perhaps play as great a role as anything in a company's decision to become involved in a new project; however, for a particular situation only one tax rate applies and this rate can usually be identified in advance.

Even with only three combinations of parameters to consider, the problem can become quite complex if several different values are applied to each of the three. Examining just three factors, each having three different values, results in 27 different combinations. It is not difficult to visualize the unmanageable proportions that such a problem could easily reach; this is why a computer is often used to calculate the effect of *all* possible outcomes whenever it is necessary to go into such detail.

The necessity for looking at all possible combinations of variables can be minimized, though. In the following example the best estimate for each parameter is used in a base case. Then one item at a time is varied with all other parameters held constant. This reduces the number of combinations to just seven—the base case plus six variations, two for each of the three parameters.

The parameters used in the seven different combinations are summarized in Table 16. All seven situations assume $100,000 in direct costs, $40,000 in indirect costs, and a 50-percent income tax rate. The base case—representing the best estimate of each parameter—shows a total sales revenue of $200,000, a capital investment of $100,000, and an 8-year economic life. After carrying out the calculations described in the project evaluation program (Table 15), the base case is found to have a ROI of 32.5 percent.

Variations 1 and 2 examine the sensitivity of the ROI to changes in the net operating profit. Variation 1 assumes a 50-percent decrease in the net operating profit, while variation 2 assumes a 50-percent increase. (The 50-percent change in net operating profit is brought about by just a $30,000 change in sales revenues.) Their respective returns are 13.5 and 49.5 percent.

Variations 3 and 4 assume changes in the capital investment requirement. Variation 3, with a 50-percent lower investment, shows a net return of 65 percent. Variation 4, requiring 50 percent more capital than the base case, has a 20.5-percent ROI.

Variations 5 and 6 decrease and increase the estimate of economic project life by half. The shorter life (4 years) lowers the ROI to 25.0 percent, while the longer life (12 years) raises the return to 34.0 percent.

The results of this sensitivity analysis are summarized in Table 17 which shows the percentage change in the project's ROI brought about by a 50-percent change in either direction for each of the three parameters being examined.

In the preceding example it is apparent that errors in forecasting the economic life of a project have relatively little effect on its apparent profitability; a high-side error is far preferable to a low-side error since it has far less effect on the project's ROI (see variations 5 and 6 in Table 16).

The ROI is especially sensitive to variations in capital requirements, with high-side estimates again preferred over estimates on the low side. The same is true to a lesser extent with net operating profit.

Under these conditions it can be concluded that it is better (in terms of overall accuracy) to overestimate than to underestimate the values of the various parameters involved in a project evaluation.

Table 16 Effect of Variables on Investment Profitability

Line	Parameter	Base Case	Variation 1	Variation 2	Variation 3	Variation 4	Variation 5	Variation 6
1	Average annual sales	$200,000	$170,000	$230,000	$200,000	$200,000	$200,000	$200,000
2	Direct production costs	100,000	100,000	100,000	100,000	100,000	100,000	100,000
3	Indirect and overhead costs	40,000	40,000	40,000	40,000	40,000	40,000	40,000
4	Net investment	100,000	100,000	100,000	50,000	150,000	100,000	100,000
5	Economic project life (years)	8	8	8	8	8	4	12
6	Income tax rate (%)	50	50	50	50	50	50	50
12	NCF	$36,250	$21,250	$51,250	$33,125	$39,375	$42,500	$34,165
13	Capital recovery rate	0.3625	0.2125	0.5125	0.6625	0.2625	0.4250	0.3417
14	ROI (%)	32.5	13.5	49.5	65.0	20.5	25.0	34.0

Table 17 Sensitivity of ROI to Parameter Changes

Variation	Description	ROI	Percent Change from Base Case
	Base case	32.5	
1	50% decrease in net operating profit	13.5	– 58.5
2	50% increase in net operating profit	49.5	+ 52.4
3	50% decrease in capital investment	65.0	+ 100.0
4	50% increase in capital investment	20.5	– 36.9
5	50% decrease in economic life	25.0	– 23.0
6	50% increase in economic life	34.0	+ 4.6

PROBABILITY AND PROJECT EVALUATION

Another important aspect of the project evaluation problem—one that is particularly significant in the predesign or early design phases of a new or proposed project—is the degree of certainty associated with the various estimates used in evaluating the project.

In the sensitivity analysis it was concluded that, of the parameters studied, ROI is least sensitive to changes in a project's economic life but is greatly affected by changes in the capital investment requirement and in the expected level of net operating profits.

In addition to knowing how much the ROI is affected by a parameter change, it is important to know the likelihood of that change occurring. When the amount and probability of change are known, an expected value can be determined and used as the basis for decision making.

The expected ROI is defined (loosely) as the sum of the products of the ROIs that will result from each possible set of conditions and the probability of each possible set of conditions occurring.

For example, if there is a 20-percent chance that a project will achieve a 30-percent ROI, a 50-percent chance of its earning 20 percent, and a 30-percent chance of realizing a 10-percent return, the expected ROI is:

$$
\begin{aligned}
0.2 \times 30 &= 6.0 \\
0.5 \times 20 &= 10.0 \\
0.3 \times 10 &= \underline{3.0} \\
& 19.0 \text{ percent}
\end{aligned}
$$

Computing the expected ROI for a new project, then, simply requires that probabilities be assigned to each possible combination of likely parameters; then that the result of each combination be weighted by the probability of its occurrence. The weighted average of all possible combinations represents the expected ROI.

In the project evaluation program discussed in the last two sections, six parameters were employed in estimating the ROI on a new project:

1. Sales revenues.
2. Direct costs.
3. Indirect costs.

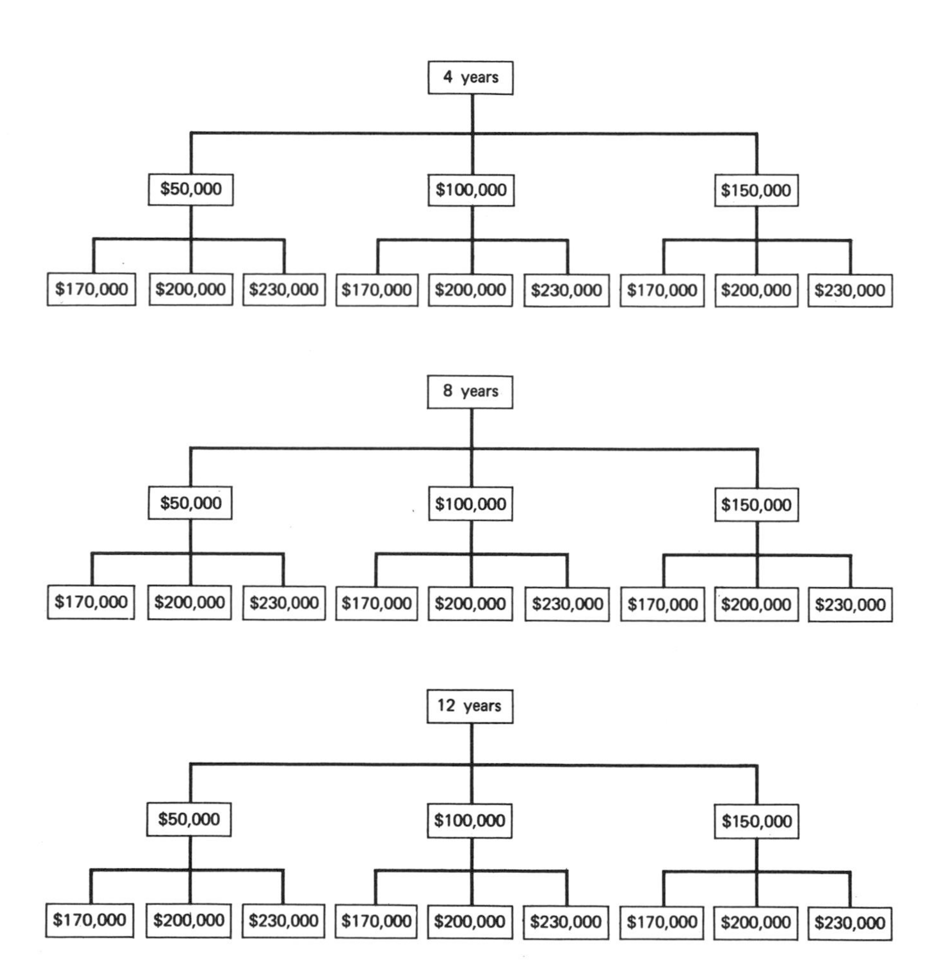

Fig. 16. *Possible combinations of three variables at three levels.*

4. Capital investment.
5. Economic life.
6. Income tax rate.

The first three items were for convenience combined into a single item—the net operating profit before deductions for depreciation and taxes. By varying just the sales revenue, then, any desired level of operating profit can be obtained. Also, since only one income tax rate applies to a given project, there is little point in introducing several different values for it.

This leaves but three factors to vary: operating profit (via changes in the sales revenues), capital investment requirement, and economic project life. If three different values were assigned to each of these three items—say a low, median, and high estimate, or an optimistic, pessimistic, and most likely estimate—the resulting 27 different combinations would be as shown in Figure 16.

In Figure 16 each of the three specified project lives is broken down into three levels of capital investment; each level of capital investment is in turn associated with three different levels of sales revenues.

Determining the return on investment associated with each of the 27 combinations shown in Figure 16 is a simple enough task (employing the project evaluation program), although it involves a considerable amount of laborious calculation. The returns thus calculated range from a low of −4.0 percent (for a 4-year project having an investment requirement of $150,000 and generating sales of $170,000 annually) to a high of 95.8 percent (for an 8-year project with a $50,000 investment and $230,000 in annual sales).

Identifying the probability of occurrence of each possible outcome, too, is quite simple, requiring only that the probability of each of the three elements— economic life, investment, and sales—be multiplied together. For example, if the probability of achieving the low, median, and high estimates for the three parameters were as follows,

	Low	Median	High
Economic life	0.1	0.7	0.2
Capital investment	0.2	0.5	0.3
Sales revenue	0.4	0.4	0.2

then the probability of a combination consisting of a median economic life, a high capital investment, and a low sales revenue would be $0.7 \times 0.3 \times 0.4 = 0.084$, or 8.4 percent. Similarly, a high economic life, a low capital investment, and a high sales revenue are expected to occur only $0.2 \times 0.2 \times 0.2 = 0.008$, or 0.8 percent of the time. The most likely combination, a median value of each parameter, has a probability of $0.7 \times 0.5 \times 0.4 = 0.140$, or

14.0 percent. The sum of the probabilities of all 27 combinations is of course 1.00, or 100 percent, since it is assumed that all possibilities are covered.

By calculating the probability of occurrence of each of the 27 possible combinations shown in Figure 16, and then by multiplying the ROI associated with each combination by its probability, the expected ROI of the project as a whole can be calculated. This is done in Table 18.

Table 18 Calculation of Expected ROI for New Project

Economic Life (years)	Capital Investment ($)	Sales Revenues ($)	ROI (%)	Probability of Occurrence	Expected ROI (%)
4	50,000	170,000	25.0	0.008	0.200
4	50,000	200,000	62.0	0.008	0.496
4	50,000	230,000	95.5	0.004	0.382
4	100,000	170,000	4.0	0.020	0.080
4	100,000	200,000	25.0	0.020	0.500
4	100,000	230,000	44.0	0.010	0.440
4	150,000	170,000	(4.0)	0.012	(0.048)
4	150,000	200,000	11.5	0.012	0.138
4	150,000	230,000	25.0	0.006	0.150
8	50,000	170,000	32.5	0.056	1.820
8	50,000	200,000	65.0	0.056	3.640
8	50,000	230,000	95.8	0.028	2.682
8	100,000	170,000	13.5	0.140	1.890
8	100,000	200,000	32.5	0.140	4.550
8	100,000	230,000	49.5	0.070	3.465
8	150,000	170,000	6.5	0.084	0.546
8	150,000	200,000	20.5	0.084	1.722
8	150,000	230,000	32.5	0.042	1.365
12	50,000	170,000	33.0	0.016	0.528
12	50,000	200,000	64.2	0.016	1.027
12	50,000	230,000	94.2	0.008	0.754
12	100,000	170,000	16.0	0.040	0.640
12	100,000	200,000	33.0	0.040	1.320
12	100,000	230,000	48.5	0.020	0.970
12	150,000	170,000	9.0	0.024	0.216
12	150,000	200,000	22.0	0.024	0.528
12	150,000	230,000	33.0	0.012	0.396
				1.000	30.397

By approaching the problem in this manner, the chances of achieving any specified ROI, or the probability that the ROI will fall within a given range, can be easily calculated. The expected ROI in this case turns out to be 30.4 percent. There is a 95-percent probability that the ROI will fall between 4.0 and 95.0 percent, and a 55-percent chance of its being between 20 and 50 percent. The probability of the project's earning *at least* 20 percent is 67 percent.

These types of observations can be made quite easily by plotting the cumulative probability distribution of the various outcomes, as shown in Figure 17. Here the probability of attaining at least a specified ROI is plotted from the data in Table 18.

To find the probability of earning between 30 and 40 precent from the graph, it is first necessary to find the probabilities of attaining these minimum returns. The graph indicates a 40-percent chance of earning at least 30 percent, and a 22.5-percent chance of reaching 40 percent. Thus the probability of earning between 30 and 40 percent is 40.0 − 22.5, or 17.5 percent. Sixty percent of the possible outcomes will fall below 30 percent, while 22.5 percent will fall above 40 percent, leaving 17.5 percent in the 30–40 percent range.

To avoid the tedious calculations involved in such an approach, a simpler method can be substituted with satisfactory results in most cases. Instead of calculating the expected ROI associated with each possible combination of

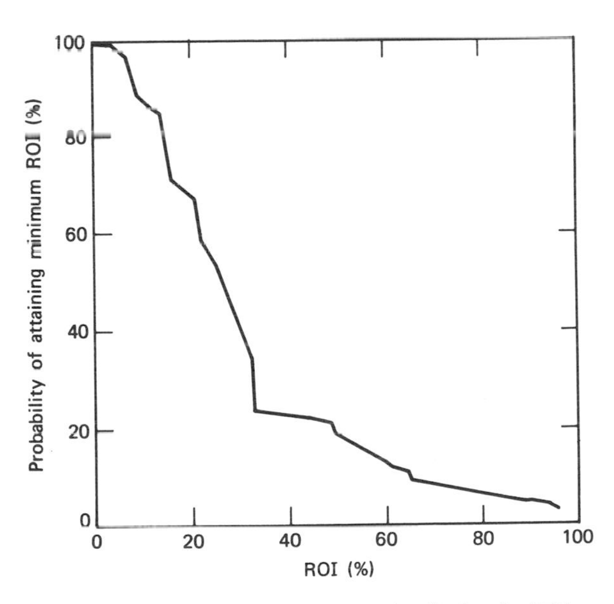

Fig. 17. *Cumulative probability distribution for ROIs.*

parameters, the ROI associated with the expected parameters can be used as a single index of profitability.

In the example the expected economic life can be calculated by multiplying the low, median, and high estimates by their respective probabilities: 0.1×4 years $+ 0.7 \times 8$ years $+ 0.2 \times 12$ years $= 8.4$ years expected life. In the same way the expected capital investment is found to be $105,000, and the expected operating profit is $194,000. Putting these expected values together (as shown in Table 19) results in an expected annual net cash flow of $33,250, representing a capital recovery rate of 0.317. The ROI associated with this capital recovery rate over the expected 8.4-year period is approximately 27.5 percent, which can be considered the expected profitability of the proposed investment. This compares with the 30.4-percent return found by the more detailed procedure.

While this approach does not identify the range of values possible in a given situation and does not identify the likelihood of achieving various levels of profitability, it does give a quick, convenient, and realistic index of project desirability.

ECONOMIC PROJECT LIFE

Most feasibility studies involve a present expenditure which is made to achieve either earnings or savings over several years in the future. This situation is encountered both in connection with new construction and with repair-replacement decisions on existing facilities. In all cases the basic question is, "How much are the future earnings (or savings) worth?" Previous sections of this chapter discussed the various factors that affect a project's ROI. The two most important factors in determining the ROI of a prospective investment are: (1) the operating profits generated by the investment, and (2) the capital investment required.

Economic life, while having somewhat less of an impact on ROI than either of these factors, is still an important consideration in choosing between alternative investment opportunities.

Importance of Economic Life

Selection of an appropriate economic life for a proposed project is usually onc of thc most difficult, and frequently one of the most questionable, aspects of the feasibility study. For municipal or public works projects, the study period may be defined as the period over which the project is to be financed; the project's economic life, then, is assumed to be equal to the life of the bonds.

Table 19 Return on Investment Calculations under Selected Conditions

Line		Expected Conditions	Worst Possible Conditions	Most Likely Conditions	Best Possible Conditions
1	Annual sales ($)	194,000	170,000	200,000	230,000
2	Direct costs ($)	100,000	100,000	100,000	100,000
3	Indirect costs ($)	40,000	40,000	40,000	40,000
4	Net investment ($)	105,000	150,000	100,000	50,000
5	Economic life (years)	8.4	4	8	8
6	Income tax rate (%)	50	50	50	50
7	Average depreciation ($)	12,500	37,500	12,500	6,250
8	Total deductions ($)	152,500	177,500	152,500	146,250
9	Net profit before taxes ($)	41,500	(7,500)	47,500	83,750
10	Income taxes ($)	20,750	(3,750)	23,750	41,875
11	Net profit after taxes ($)	20,750	(3,750)	23,750	41,875
12	Net cash flow ($)	33,250	33,750	36,250	48,125
13	Capital recovery rate	0.317	0.225	0.363	0.963
14	ROI (%)	27.5	–4.0*	32.5	95.8

*Any capital recovery rate of less than 1/N (where N is the economic life in years) does not recover the entire investment over the project's life and therefore represents a negative ROI. A capital recovery rate of 1/N corresponds to a zero percent return; any rate of more than 1/N indicates a positive return.

For industrial projects, though, the study period probably is much shorter than the life of the equipment and facilities under consideration. Seldom does an industrial firm go beyond a 10-year operating period in analyzing a new venture, and 5- to 8-year study periods are common. Actually, what happens beyond 10 years makes very little difference in most industrial feasibility studies; but whether the study period is 5 years or 10 years can have a very substantial impact on the apparent ROI.

Effect of Economic Life

Take, for example, a company in a 50-percent tax bracket considering an investment of $1 million in a project expected to earn $400,000 annually before taxes. If it is assumed that the investment is such that it can be written off (depreciated) over the project's economic life, the depreciation charge is deducted from the $400,000 to obtain the net taxable income, and cash flow is calculated as the net profit after taxes plus the depreciation charge. Over a 5-year period, then, depreciation will be $200,000 per year and the net cash flow will be (400,000 − 200,000)(0.50) + 200,000 = $300,000. A $300,000 net cash inflow for 5 years resulting from a $1 million initial investment is equivalent to 15.2 percent ROI.

Going to a 10-year economic life would result in a net annual cash inflow of (400,000 − 100,000) (0.50) + 100,000 = $250,000. This cash inflow over a 10-year period is equivalent to a 21.4-percent return.

Extending the project's economic life to 15 years, though, brings about relatively little change in the ROI. Over a 15-year period an average net annual cash flow of $233,000 will be realized, resulting in a yield of 22.1 percent on the invested capital. This is the same yield that would be obtained over a 20-year period.

The top curve in Figure 18 illustrates the effect of economic life on the expected rate of return in the situation just described. An unusual feature of the relationship is that as the estimated life increases beyond a certain point —about 17 years in the example—the apparent ROI actually declines. Thus a 17-year life brings a 22.2-percent return; but with an infinite life, annual depreciation will be zero, and net cash flow will be a steady $200,000, corresponding to an even 20-percent ROI.

There is, in every instance in which income taxes are involved, some intermediate time period at which the return on investment is a maximum; this point might be termed the "optimum economic life."

The two lower curves in Figure 18—representing gross annual returns of 30 and 20 percent of the invested capital—show the same general relationships as the previous example, although the ROI is effected more gradually by the project's economic life at these lower levels.

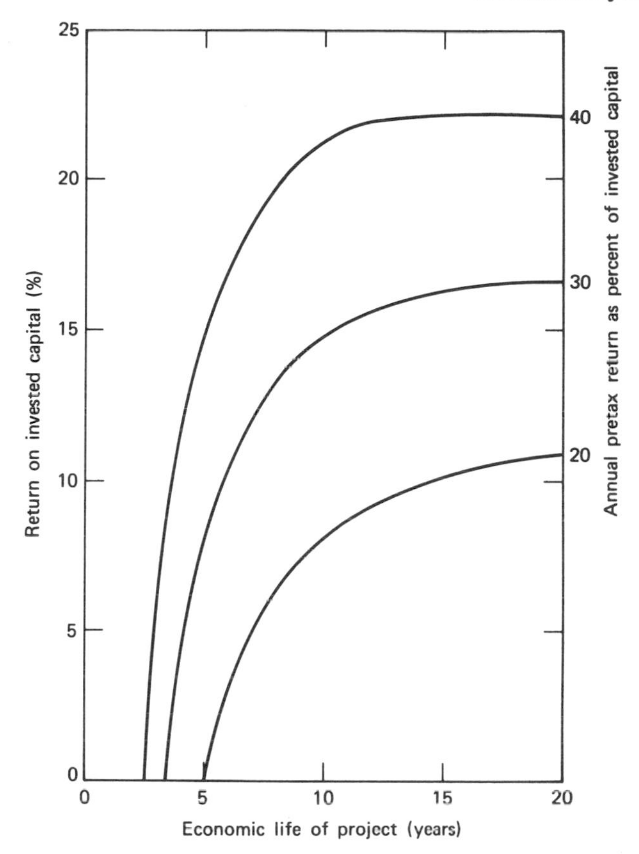

Fig. 18. *Profitability versus economic life for depreciable investments (50-percent tax bracket).*

With the $200,000 annual return on a $1 million investment (the bottom curve), the project will just break even at 5 years. At a 10-year life it will show a return of 8.1 precent; at 15 years, 10.2 percent; and at 20 years, 10.9 percent. At 20 years it will not yet have reached its optimum economic life, but on extending the life indefinitely, the ROI will gradually decline to 10.0 percent.

The same general relationships hold in the middle curve of Figure 18, representing a 30-percent gross annual return on the invested capital. Here the ROI will eventually level off at 15.0 percent but will reach a maximum of about 17 percent somewhere in the 10–25-year area.

These three examples are all based on a 50-percent tax rate, and an investment that can be fully depreciated over the project's economic life, with a uniform annual series of cash returns and with depreciation computed on a straight-line basis. Employing accelerated depreciation methods would increase the apparent ROI by a percent or two but would not alter the general relationships.

Tax-Free Investments

For a tax-free investor, though, depreciation has no effect on cash flow, and the net cash flow is the same as the gross annual return. In this case, the ROI increases as the project life increases. A 10-percent per year net return on the invested capital will break even in 10 years; it will earn 5.6 percent over 15 years, 7.8 percent over 20 years, 9.3 percent over 30 years, 9.8 percent over 40 years, and approach 10.0 percent as the time period is lengthened infinitely. In this case there is no "optimum" economic life in terms of the return on investment. Figure 19 illustrates the relationship between ROI and net annual returns in a tax-free situation.

Some of the main distinctions that must be made when conducting feasibility studies for industrial versus public clients are:

1. Industrial projects must consider income taxes, to which public projects are not subjected.

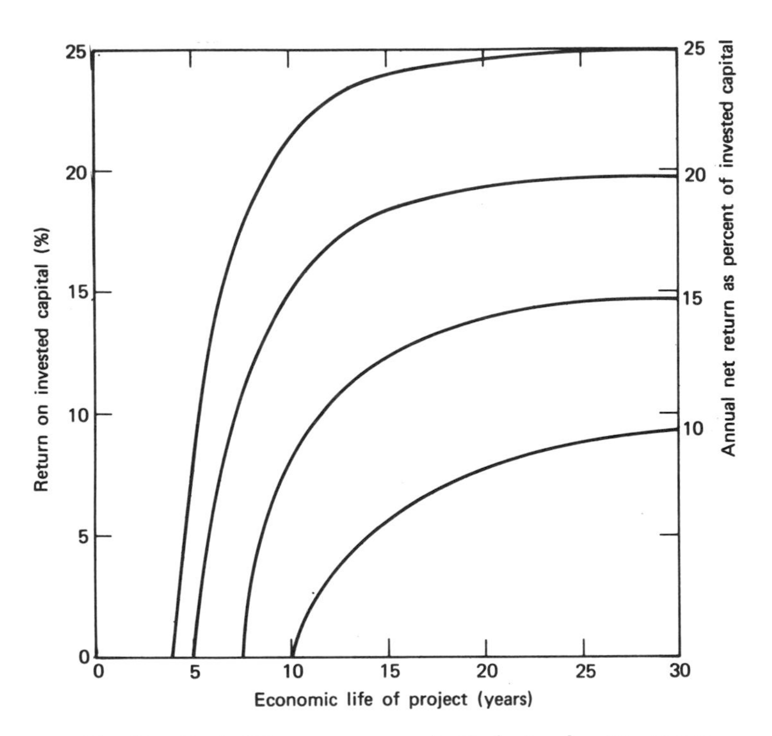

Fig. 19. *Profitability versus economic life for tax-free investments.*

2. Because of the income tax structure, the choice of an appropriate depreciation method is important in industrial studies, while depreciation is not an important consideration (in an accounting sense) in studies for public clients.

3. Most industrial firms require a return on invested capital above 15 percent, while most public works projects are concerned primarily with covering their cost of capital—usually less than 6 percent.

4. For industrial projects a relatively short payout period is usually necessary, while for public projects it is advantageous to extend the project life over as long a period as possible.

5. Because of all these factors, industrial projects have some identifiable optimum economic life; extending the study period beyond this point reduces the apparent ROI.

INCOME TAXES

Income tax rates similar to interest rates, have become an important tool for the government in its attempts to control inflation. A high income tax rate decreases the supply of money available for business investment and consumer spending, while high interest rates discourage borrowing (or debt financing) and make new investment appear less desirable from an economic standpoint.

Consequently, income taxes are an important part of most feasibility studies for industrial clients, while interest rates may be a controlling factor for public works and other nontaxpaying entities. Privately owned utilities are seriously affected by both tax rates and interest rates, since they usually rely heavily on long-term borrowing to finance new projects.

In the *which* type of feasibility study—that is, studies in which the objective is to identify *which* of several alternatives is most economically attractive—income taxes have little effect on the outcome, since the same tax rate applies to all alternatives. In this situation whatever is the most attractive before taxes is likely to be the most attractive after taxes also.

But in the *whether* type of feasibility study—in which the question is not which way to do it but *whether* to do it at all—income taxes can have a major impact on the investment decision. For example, if a company in a 50-percent bracket requires that new projects yield a minimum return on investment of 15 percent, then the pretax return will have to be in the 30-percent range for the investment to qualify as a satisfactory commitment.

If tax-exempt municipal bonds are currently yielding 5.5 percent with virtually no risk, a commercial bank's reluctance to loan money even to its prime customers at anything less than 11.0 percent would certainly be understandable.

High income tax rates, then, discourage business investment, while low tax rates encourage investment—essentially the same effect as high or low interest rates.

Income Taxes and Investment Economics

In comparing alternative investment opportunities, different cash flows between the alternatives are changed by the amount of the actual tax payable on the differences in accounting profits. Regardless of whether the taxes are assessed on ordinary income or capital gains, they can be considered simply another cash expense or outflow so long as they are calculated according to appropriate legal and accounting standards.

The effect of income taxes on the apparent attractiveness of a proposed investment can be visualized from the simple example shown in Table 20. This example shows an initial $1000 investment, recovered through a depreciation charge of $200 annually for 5 years, generating additional revenues of $400 per year over that period. The net profit before taxes, then, is $200

Table 20 Effect of Income Tax Rate on ROI

Year	Gross Cash Flow	Depreciation	Net Profit before Income Tax	Net Cash Flow after Income Taxes at Various Tax Rates 0%	30%	50%	100%
0	(1000)	0	0	(1000)	(1000)	(1000)	(1000)
1	400	200	200	400	340	300	200
2	400	200	200	400	340	300	200
3	400	200	200	400	340	300	200
4	400	200	200	400	340	300	200
5	400	200	200	400	340	300	200
ROI:				28.7%	20.7%	15.3%	0%

each year, and the corresponding gross cash flow is $400. Net cash flow of course decreases as the tax rate increases, ranging from $400 at a zero tax rate, to $300 at a 50-percent rate, and on down to $200 at a 100-percent tax rate.

Return on Investment

The ROI associated with these tax rates has a maximum of about 29 percent at the zero rate and a minimum of zero at the 100-percent tax rate. The ROI resulting from intermediate tax rates can be interpolated roughly as 15 percent at a 50-percent rate, 18 percent at a 40-percent rate, and 20 percent at a 30-percent rate. In this example each 10-percent increment in tax rate changes the ROI by about 3 percent.

At a zero tax rate cash flow is the same regardless of the depreciation method employed. But when income taxes are considered, a slightly higher ROI can be obtained by using an accelerated depreciation method rather than the straight line method.

Other Effects of Income Taxes

Income taxes may play an important role in setting financial policy for a company. Since interest payments are not taxed, debt financing has been preferred to equity financing by many established companies unless equity capital can be obtained by giving up relatively little in present earnings.

Even personal income taxes can affect, indirectly, a company's cash position, although personal taxes usually do not influence investment policies concerning new projects. In attracting new equity capital, outside investors normally prefer capital gains to dividends because of the difference in tax liabilities; consequently, if the company can reinvest its surplus cash to generate earnings growth and resulting capital appreciation, investors will usually respond by placing a higher value on the company's equities. Thus the cost of equity capital may be lowered substantially and, with a lower cost of capital, more projects may be considered economically feasible.

Regardless of the effects of income taxes on a company's operations, it is essential that they be given proper consideration in the feasibility study; this requires a knowledge of income tax laws and regulations, and a familiarity with the company's tax status.

SUMMARY

The DCF technique offers a logical and generally accepted means of relating the cash earning potential of a proposed project to its investment requirements. The project's ROI thus measured defines the maximum interest rate at which money can be borrowed to finance the project and subsequently repaid out of cash earnings over the project's economic life. Simple DCF problems may be solved mathematically, although most are far too complex for a mathematical solution. A trial-and-error approach is generally used, usually in conjunction with either algebraic or graphical interpolation. Some of the difficulties associated with DCF analysis can be circumvented by employing graphical techniques. The *investment profit prophet* for example, can be applied to projects having an economic life of from 5 to 15 years and yielding a return between zero and 40 percent. For projects assumed to have uniform cash flows, a simplified approach can be taken; this approach, incorporating just six factors, is especially useful in screening new projects before all the details are fully known. By varying the six factors systematically, the sensitivity of the project's ROI to each factor can be easily estimated. By assigning probabilities to each of the factors, the expected ROI for the project can be calculated, along with the overall range of possible outcomes. Of the six major factors, the capital investment requirement and the gross operating profits are generally found to have the greatest effect on the ROI of a given project. Economic life, however, may be an important factor in comparing different projects to determine which is most desirable; and income taxes exert a major influence on a decision regarding whether or not a project will even be considered.

5

Measuring Investment Performance

One of the important differences between a consulting engineer and an engineer in industry is apparent in the follow-up of a recommended investment. While the engineer in industry may have to live with his decisions and recommendations for many years (no matter how embarrassing they may be), the consulting engineer may never even find out whether or not his recommendations were followed. If they were followed, he will probably never know just how closely the results of his recommended projects came to the estimates and forecasts upon which they were based.

But unless a project is followed up, there is little the engineer can do to improve his performance in project evaluation, planning, and forecasting. Incentive is lacking to develop new or better techniques when the results of existing techniques are not known.

This is where the postinvestment appraisal comes in. The postinvestment appraisal is simply a comparison of the actual results of an investment decision with the results originally estimated in justifying the investment decision in the first place. Thus the postinvestment appraisal offers the engineer an opportunity to objectively appraise the results of his own work by showing to what extent investment proposals have achieved the results forecast for them.

Other equally descriptive terms sometimes used for the postinvestment appraisal include postinvestment audits, postcompletion reviews or audits, performance reviews, experience reports, or postmortems.

PURPOSES OF THE POSTINVESTMENT AUDIT

The characteristics of a postinvestment audit depend largely upon the purposes for which it is made. For many projects the engineer's first "audit" occurs when bids are opened for the initial construction work. At this point a hasty review of the entire investment proposal may be called for. Many unpleasant surprises can be avoided by some extra care during the construction phase of a new project.

Checking the accuracy of a capital cost estimate, though, is only one of many useful purposes to which postinvestment appraisals can be put. Other equally important reasons for making these audits include:

- Improving the evaluation of future projects.
- Insuring realistic estimates by sponsors of new projects.
- Improving the operating performance of completed projects.
- Providing management with information regarding the success of new projects.
- Identifying areas where improvements are needed, or where remedial action can or should be taken.
- Determining the soundness of original assumptions.
- Disclosing inadequacies in the form or content of requests for capital appropriations.

A successful postinvestment appraisal program must be carried out on a very impersonal basis, exercising great care to avoid placing unwarranted blame on specific groups or individuals. So many different events and decisions may have preceded the investment decision that it is extremely difficult to pinpoint the specific factors that affected its ultimate profitability.

Properly administered, and used in a systematic and thorough manner, postinvestment appraisals can greatly enhance the engineer's knowledge and understanding of what has happened in the past and thus improve his capabilities for performing under similar conditions in the future.

APPLICATIONS OF THE POSTINVESTMENT APPRAISAL

Since the consulting engineer's responsibility for a new project generally ends shortly after the startup period is over, his interest in the project may unfortunately end there too. As pointed out earlier, this is one of the main differences between the consultant and his client.

According to surveys of major United States industrial firms, nearly 90 percent of the companies employ performance reviews on selected new projects, beyond just an accounting review of actual versus estimated capital expenditures.

The best type of project to be subjected to postauditing is the kind that has been entered into directly for profit, in which both operational and financial performance can be measured. For private or industrial firms, typical projects that should be postaudited include new product lines, replacement projects, and expansion projects. For municipal or public organizations, postinvestment appraisals are particularly important on projects financed through revenue bonds, service charges, or other revenue sources specifically related to the project.

But regardless of the type of project or client involved, the key to obtaining maximum benefits from the postinvestment appraisal is to check out the most important factors involved in initially justifying the project, especially those factors to which the project's economic success is most sensitive. A project's ROI, for example, has been shown to be most sensitive to its initial capital outlay and its gross operating profits. These, then, are the logical factors to scrutinize in any postinvestment reviews of projects entered into for the purpose of earning an ROI.

PROBLEMS AND LIMITATIONS

The industrial firm, monitoring the self-inflicted results of its own decisions on its own terms, is faced with several problems in administering a postinvestment appraisal program. The consulting engineer, in addition to having all the conventional problems, is also apt to be faced with some problems in gaining access to the data necessary for a realistic audit.

One of the main problems from the engineer's standpoint is that while the audit measures the results of what was done it does not indicate whether or not the original design was the best possible. For example, an audit of a sewage treatment facility might show that the trickling filter design criteria were correct, the construction costs were accurate, the operating costs came out as expected, the organic and hydraulic loadings were properly anticipated, and the effluent quality was predicted. But the audit does not reveal whether the trickling filter plant that was installed was preferable to whatever alternatives were available at the time the investment decision was made.

Internal problems abound within companies or departments whose investment programs are being aduited. Corporate "in-fighting," a general lack of incentive, and extreme reluctance to revive old problems that would sooner be forgotten pose serious barriers to the development of good postinvestment audits. For a variety of practical reasons, it is quite difficult to come up with good investment reviews, regardless of how valuable they might be.

Some of the other problems associated with postinvestment appraisals include a tendency toward complacency when performance is good and a corresponding reluctance to expose the facts when performance is poor. Also, obtaining the appropriate financial data may be a problem in that the data in a company's accounting system may not be directly indentifiable with specific projects, and some costs such as overheads and administrative expenses may not be easily or rationally allocable. Obtaining relevant operating data, especially from municipalities, may turn out to be strictly a do-it-yourself project for the engineer.

ADMINISTRATION OF THE PROGRAM

There are four major considerations in the administration of an effective post-investment appraisal program. They are:

1. Assignment of responsibility.
2. Selection of projects.
3. Timing of audits.
4. Reporting audit results.

RESPONSIBILITY. Responsibility for conducting an audit of a completed project depends primarily upon the reasons for performing it. If the main purpose of the audit is to help insure realistic estimates by checking up on the accuracy of the firm's estimators, then responsibility should be assigned to a group operating independently of those who prepared the original estimates. But if the purpose of the audit is to improve future evaluation techniques, then the same people who prepared the estimates should also conduct the audit.

SELECTION OF PROJECTS. The selection of projects to be subjected to review is usually determined by some set dollar amount, with the range extending from a minimum of anywhere between $5000 and $100,000. The projects selected should be those expected to yield the greatest amount of information for use on similar future projects.

TIMING. Timing of the postinvestment appraisal is extremely critical if the best possible results are to be obtained. If the performance review is conducted too soon following the expenditure, there will be no accurate reflection of the project's long-range success or failure. But if the audit comes too late, the data are harder to obtain and the potential benefits of the review may be lessened because of changes in personnel and procedures. From the consultant's viewpoint calling on a client to review an unsuccessful project's results may prove an embarrassing experience. Nevertheless, if embarrassing experiences are to be avoided in the future, it is necessary to learn what went wrong in the past and how the same mistakes might be avoided. However, a close working relationship with a client on a good, successful project can be a pleasant and valuable experience for all, leading to continually improving approaches to new problems.

AUDIT REPORTS. A typical accomplishment or performance report on a completed project requires a flexible format but usually consists of two main sections. The first section includes a review of the original objectives of the project, general comments regarding the conditons encountered during the time period over which the review was made, and a brief statement concerning the outlook for the future. The first section of the performance report, then, is primarily narrative and subjective in nature. The second section contains figures on profit performance, or comparisons of specific estimates of sales, manufacturing or production costs, volume handled, or whatever other parameters were used in the initial study, along with comparable data from the originally proposed project. Adjustments should be made in the original estimates to reflect changes in wage rates, materials prices, and general economic conditions.

INTERPRETATION OF RESULTS

After all the information has been collected that will make possible a meaningful comparison of the actual results of a project with the estimates upon which the project was originally based, the next step is to interpret any significant differences between actual and estimated results. In doing this some distinction should be made between two different types of estimates:

1. Estimates based on formal forecasting methods.
2. Estimates based on individual judgment.

If the estimates were based on formal techniques developed from past experience, it is relatively easy to verify the validity of the techniques simply by studying a sample of the estimates they produced. It is almost equally easy to determine whether the forecasting method was properly applied in a specific case.

When estimates were based on an individual's judgment, though, there is little to be learned by comparing an actual result with an estimate in a specific situation. More can be learned in this case by comparing the factors that were considered important in making the original estimate with the factors that actually influenced the project's performance—essentially an after-the-fact sensitivity analysis. If an individual's judgment is to be appraised, he can be viewed as a "black box" by comparing the actual and estimated results on several projects in which his judgment was exercised.

Whenever possible, interpretation of the comparisons between actual and estimated results should be based on groups of similar projects rather than upon single projects. If the purpose of the postinvestment audit is to improve forecasting methods or to insure more realistic estimates, considering more projects will provide better results.

In all comparisons the estimates and the bases on which they were made should be explicitly defined and available at the time of the audit. All factors considered in the judgments made in connection with the estimates should be clearly stated, and all forecasts and assumptions used in the final estimate should be recorded. Only then is it be possible to determine which of the preliminary forecasts or assumptions caused the final forecast to vary from the actual results.

It is apparent that economic justification of capital expenditures encompasses two broad areas. First, it requires the development and application of reliable methods for determining the financial and/or operating success of new capital investments—methods for screening new projects and for evaluating proposed investments. Second, it is concerned with a continuing review of the investment decision process through comparisons of actual results with the assumptions and data upon which the expenditures were initially justified.

CASE HISTORIES

The following three case histories have been drawn from the experience of manufacturing firms in different areas and involve different types of projects. The first two examples deal with new manufacturing facilities built to produce new products; the third example describes the results of an acquisition.

While these three examples are not spectacular, they at least demonstrate the usefulness of postinvestment analyses. In most cases it is found that actual profitability is lower than predicted in the preinvestment evaluation of a new venture. The cost engineer may be overly optimistic in assuming that nothing will go wrong even though he knows that something always does. Some of the most common causes of error in investment evaluations are:

- A new product is given too high a selling price.
- The enthusiasm of the market for the product is overestimated, and an unattainably high sales volume is projected.
- The time required to penetrate new markets is underestimated.
- Reactions of competitors are not correctly anticipated.

Realistic assessments of markets, prices, and production costs—entailing in-depth market and cost research—are necessary components of a good investment analysis, and the cost engineer must guard against being overly optimistic despite the enthusiasm shown by a new project's sponsors. Investment or capital requirements are the most reliable part of most evaluations, since the cost engineer or estimator has generally had enough experience in that area to develop reasonably accurate figures. Market research, however, is a field calling for special qualifications in areas in which most engineers are less experienced.

Example 1: Facilities to Manufacture a New Chemical Product

This project dealt with a new development that was discovered by the company's research department. Because of the nature of the product and a high level of management enthusiasm for it, the company felt that the product should be rushed into production as rapidly as possible. The cost estimates used to justify the venture were made without full design information and without a comprehensive market analysis, as evidenced by a comparison of actual with estimated costs; the values actually realized in this example show little relationship to those estimated. But in this case nobody cared because the project turned out to be so successful. The cost engineer's function was simply to come up with some numbers quickly, using whatever short-cut techniques he could muster. Table 21 summarizes the results.

As it turned out, the new chemical product was much better than any competitive products on the market at the time, and the first-year sales exceeded the estimates by 20 percent. This, coupled with a fixed investment that was nearly 30 percent below the estimate, resulted in a percentage return of 29 percent, almost double the estimated return of 15 percent.

This example is unusual in that everything turned out better than anticipated. Since a new venture must show a reasonably attractive return in the preinvestment analysis before it is approved, cost estimating errors on the high side (giving lower profits and percentage returns) are not usually discovered; the projects are never undertaken. Consequently, the postinvestment

Table 21 Experience Report on Facilities to Manufacture a New Chemical Product

	Estimated ($)	Actual ($)	Ratio, Actual to Estimated
Total capital requirement	1,500,000	1,057,000	0.71
Net sales	1,200,000	1,444,000	1.20
Costs			
Production costs	300,000	361,000	1.20
General overheads	210,000	289,000	1.38
Fixed charges	240,000	179,000	0.75
Total Costs	750,000	829,000	1.11
Net profit before taxes	450,000	615,000	1.37
Income taxes	225,000	302,000	1.34
Net profit after taxes	225,000	313,000	1.39
Annual return on capital	15.0%	29.6%	1.97

analysis usually reveals projects that appeared to be economically attractive in the initial analysis but performed less well than expected.

Example 2: Manufacturing Facilities for a Plastics Intermediate

This project was proposed because of the patent protection available on certain manufacturing processes. The preinvestment analysis indicated a high return on investment and a rapid payback. Table 22 shows the estimated and actual figures for the plant's first year of operation.

While the fixed investment was slightly underestimated, the chief problem here was failure to anticipate the market and the competition properly. The initial reaction of competitors was to cut prices and increase their own promotional activities, resulting in lower sales at lower prices than estimated for this new plant. The net sales actually realized were more than 40 percent below the estimated first-year sales, resulting in a percentage return of only a sixth of that projected. Other variations between actual and estimated results were of a minor nature, attributable mainly to the lower scale of operation.

Table 22 Experience Report on Manufacturing Facilities for a Plastics Intermediate

	Estimated ($)	Actual ($)	Ratio, Actual to Estimated
Total capital requirement	815,000	842,000	1.03
Net sales	1,150,000	682,000	0.59
Costs			
Production costs	650,000	416,000	0.64
General overheads	110,000	96,000	0.87
Fixed charges	115,000	126,000	1.10
Total costs	875,000	638,000	0.73
Net profit before taxes	275,000	44,000	0.16
Income taxes	140,000	21,000	0.15
Net profit after taxes	135,000	23,000	0.17
Annual return on capital	16.6%	2.7%	0.16

This example emphasizes the need for thorough market research. Even a good product in a strong market will be successful only if the company is capable of meeting its competition in terms of price, distribution, and service, and these factors should all be considered before committing capital to a new venture. A poor estimate of market strength can easily result in a financial disaster.

Example 3: Acquisition of Metal Fabricator

The experience report shown in Table 23 describes the results of an acquisition. The company whose assets were purchased in this case was a leading customer of the firm making the acquisition, and the primary purpose of the transaction was to protect a position as a raw materials supplier. This objective was attained, even though only a modest return on invested capital was realized. Thus the venture could be considered a successful acquisition; the company operated satisfactorily and performed generally as expected.

Table 23 Experience Report on Acquisition of a Metal Fabricator

	Estimated ($)	Actual ($)	Ratio, Actual to Estimated
Total capital requirement	800,000	792,000	0.99
Net sales	1,000,000	1,144,000	1.14
Costs			
Production costs	740,000	1,019,000	1.38
General overheads	85,000	134,000	1.58
Fixed charges	120,000	128,000	1.07
Total costs	945,000	1,281,000	1.36
Net profit on manufacturing operations	55,000	(137,000)	—
Profit on sales to manufacturing	85,000	245,000	2.88
Net profit before taxes	140,000	108,000	0.77
Income taxes	65,000	43,000	0.66
Net profit after taxes	75,000	55,000	0.87
Annual return on capital	9.4%	8.2%	0.87

While net income was substantially less than estimated (chiefly as a result of price declines), the total profits realized on the sale of raw materials to the fabricating operation made up for most of the deficit. So, in terms of the company's overall operations, the net profit was just 13 percent less than expected, and the ROI was off by about the same amount.

Some of the major problems encountered in studying prospective acquisitions include difficulties involved in making good estimates of operating costs, in realistically evaluating a company's market strengths and weaknesses, and in properly assessing the technical and managerial competence of people in the acquired company. These difficulties are often compounded by the reluctance of the company to reveal detailed operating data or any adverse information during the negotiations; only its good side is presented to the prospective buyer.

SUMMARY

Economic justification of capital expenditures actually involves two phases: (1) the investment evaluation, conducted prior to any capital commitment; and (2) the performance review, made after the project is in operation. The adequacy of investment analysis techniques can be measured only in terms of how well they anticipated actual investment performance. The performance review helps in improving estimating and project evaluation techniques, as well as in identifying problem areas and providing management with useful information regarding the progress of new ventures. The key considerations in conducting an effective postinvestment appraisal are: (1) assigning the right people to the job; (2) selecting the appropriate projects for review; (3) conducting the appraisal at the proper time; and (4) reporting the results of the audit in a suitable format for interpretation. In most cases actual results are less favorable than estimated, since estimates indicating poor results discourage the project's undertaking and the poor estimates are never revealed. Although it is important to guard against overoptimism in investment evaluation, the economic consequences of overlooking good projects are equally real, although less indentifiable, than losses suffered from entering unprofitable ventures. The experience reports of companies in various fields emphasize the need for, and illustrate the consequences of not having, thorough cost and market research coupled with sound estimating techniques in the evaluation of new ventures. Properly administered, postinvestment appraisals can lead to a better understanding of what has happened in the past and improve the engineer's capabilities for future performance.

6

Depreciation

Depreciation is probably one of the least understood but most important elements in economic analysis. It is essentially a legal or accounting device, and as such is necessarily artificial. The chief importance of depreciation in economic analysis is in its effect on income taxes, which subsequently is reflected in a project's cash flows.

The importance of economic life in a project's cash flows and the effect of cash flows on profitability were covered in Chapter 4. This chapter discusses various depreciation concepts, methods of estimating economic life and computing depreciation charges, and the effect of depreciation on profitability.

As in the material covered in earlier chapters, the individual's point of view makes a big difference. An accountant is primarily interested in the depreciation of assets already in service, from the standpoint of tax-deductible expenses. An engineer, however, is interested mainly in the depreciation of assets not yet acquired for use in new ventures, from the standpoint of capital recovery. Both the engineer and the accountant should recognize the role of depreciation in generating cash flows, and the impact of various depreciation schedules on profitability.

If responsibility were to be allocated between accountants and engineers, the accountant would probably be concerned with analysis and the engineer with estimation. In any event depreciation has many meanings and can be measured in many different ways.

DEPRECIATION CONCEPTS

The various definitions of depreciation can be placed in one of four categories: (1) impaired serviceableness, (2) decrease in value, (3) difference in value, or (4) amortized cost.

IMPAIRED SERVICEABLENESS. The first concept—that of impaired serviceableness—implies a loss in utility or efficiency, although not all assets necessarily depreciate in this sense. The impairment may be either physical—such as caused by wear and tear—or it may be functional, attributable to obsolescence or inadequacy. Sometimes referred to as the "engineering" concept of depreciation, this approach is often used in repair-replacement studies.

DECREASE IN VALUE. The second concept—a decrease in value—is unfortunately a commonly employed definition of depreciation which has absolutely no clear-cut meaning. It requires that the asset's value be computed at different points in time, without defining what is meant by value—an even more elusive term than depreciation. The decrease in value may refer either to market value or value to the owner, regardless of what caused the decline.

DIFFERENCE IN VALUE. The third concept, which also might be called the "appraisal" concept, refers to the difference between the present value of the old property and the present value of a hypothetical new, but similar, property. For example, many public utilities' rates are based on the "fair value" of their property, with value sometimes defined as reproduction cost new less depreciation; depreciation in this case is usually related somehow to the asset's physical condition.

AMORTIZED COST. The fourth concept of depreciation is the familiar accounting approach in which the asset's original cost is amortized, or written off, over its useful service life. Depreciation in this case is defined simply as the difference between original cost and current book value. This is the approach usually taken in feasibility studies in which proposed investments are evaluated.

The following discussion is limited to the last concept—that of amortized cost, as used in determining investment feasibility. Here the objective is, theoretically, to recover completely the original investment by charging some predetermined portion of the initial cost during each year of the accounting period.

ESTIMATING SERVICE LIFE

Obviously, the accuracy of any depreciation method is largely dependent upon the accuracy with which the asset's service life can be estimated. This estimate, along with the method used for computing depreciation (straight line, sum of years digits, or declining balance) provides the basis for a major portion of the asset's annual ownership cost. Consequently, the time period over which the study is to be made exerts a significant influence on a proposed investment's apparent profitability, feasibility, or desirability.

While in every case the entire amount of an investment can eventually be recovered through depreciation charges, it is the timing of these charges—reflecting the time value of money—that largely determines whether or not the investment is a good one. Although caution dictates that a conservative estimate of useful service life be used, especially in studies in which a relatively high degree of risk may be encountered, the ultimate objective in any service life estimate is to have the estimate fit the experience of the particular property being depreciated; the annual depreciation charge can be no more accurate than the estimate of service life.

There are several different ways to estimate the useful service life of capital assets. The most sophisticated methods involve actuarial analysis; less sophisticated, but extremely useful and far more convenient, are the Internal Revenue Service guidelines.

IRS GUIDELINES. The Internal Revenue Service Publication No. 456, *Depreciation Guidelines and Rules*, provides tables of useful lives of a variety of items, arranged by industry. Tables 24–27 summarize the depreciation guidelines for various types of assets used in business and industry.

Use of the IRS guidelines is not mandatory; they are intended solely as a guide to "what might be considered reasonably normal periods of useful life." In practice, the service lives used as the basis for depreciation accounting should be determined by the asset's particular operating conditions and experience, tempered by judgment concerning the probability of technological improvements and economic changes.

Special provisions unrelated to physical or economic life are sometimes made by the IRS for rapid amortization of certain type of assets for the purpose of stimulating business investment or encouraging investment in facilities that will be in the public interest. A 1971 regulation, for example, provided for a 5-year writeoff of approved pollution control facilities.

While depreciation deductions are generally questioned by the IRS only when there is a clear and convincing reason to do so, the periods of estimated useful life used by the taxpayer are subject to review by the IRS, and there should be some valid means of substantiating the time periods used. The most convincing proof is through actuarial analysis.

Table 24 Depreciation guidelines for Assets Used by Business in General

Type of Asset	Depreciable Life (years)
Office furniture, fixtures, machines, and equipment	10
Transportation equipment	
Aircraft	6
Automobiles	3
Buses	9
Light general-purpose trucks	4
Heavy general-purpose trucks	6
Railroad cars	15
Over-the-road tractor units	4
Trailers and trailer-mounted containers	6
Water transportation equipment	18
Land improvements	20
Buildings	
Apartments	40
Banks	50
Dwellings	45
Factories	45
Garages	45
Grain elevators	60
Hotels	40
Loft buildings	50
Machine shops	45
Office buildings	45
Stores	50
Theaters	40
Warehouses	60
Subsidiary assets	*

*To be determined according to the facts and circumstances.

ACTUARIAL ANALYSIS. Actuarial appproaches to estimating service life involve the construction of frequency curves, survivor curves, and probable life curves. These curves for physical property are similar both in derivation and use to the mortality curves prepared by insurance companies to reflect human life expectancy. The survivor curve for physical property is developed from historical data on similar property, just as the mortality curve is developed by actuarial analysis of vital statistics on humans.

Table 25 Depreciation Guidelines for Assets Used in Nonmanufacturing Activities

Type of Asset	Depreciable Life (years)
Agriculture	
Machinery and equipment	10
Animals	
Cattle, breeding or dairy	7
Horses, breeding or work	10
Hogs, breeding	3
Sheep and goats, breeding	5
Trees and vines	*
Farm buildings	25
Contract construction	
General contract construction	5
Marine contract construction	12
Fishing	*
Logging and sawmilling	
Logging	6
Sawmills	10
Portable sawmills	6
Mining	10
Recreation and amusement	10
Personal and professional services	10
Wholesale and retail trade	10

*To be determined according to the facts and circumstances.

Table 26 Depreciation Guidelines for Assets Used in Manufacturing

Type of Asset	Depreciable Life (years)
Aerospace industry	8
Apparel and fabricated textile products	9
Cement manufacture	20
Chemicals and allied products	11
Electrical equipment	
Electrical equipment	12
Electronic equipment	8
Fabricated metal products	12
Food and kindred products	12
Glass and glass products	14
Grain and grain mill products	17
Knitwear and knit products	9
Leather and leather products	11
Lumber, wood products, and furniture	10
Machinery, except electrical and metalworking	12
Metalworking machinery	12
Motor vehicles and parts	12
Paper and allied products	
Pulp and paper	16
Paper finishing and converting	12
Petroleum and natural gas	
Drilling, geophysical and field services	6
Exploration, drilling and production	14
Petroleum refining	16
Marketing	16
Plastics products	11
Primary metals	
Ferrous metals	18
Nonferrous metals	14
Printing and publishing	11
Professional, scientific, and other instruments	12
Railroad transportation equipment	12
Rubber products	14
Ship and boat building	12
Stone and clay products, except cement	15
Sugar and sugar products	18
Textile mill products, except knitwear	
Textile mill products	14
Finishing and dyeing	12
Tobacco and tobacco products	15
Vegetable oil products	18
Other manufacturing	12

Table 27 Depreciation Guidelines for Assets Used in Transportation, Communications, and Public Utilities

Type of Asset	Depreciable Life (years)
Air transport	6
Central steam production and distribution	28
Electric utilities	
Hydraulic production plant	50
Nuclear production plant	20
Steam production plant	28
Transmission and distribution facilities	30
Gas utilities	
Distribution facilities	35
Manufactured gas production plant	30
Natural gas production plant	14
Trunk pipelines and related storage facilities	22
Motor transport—freight	8
Motor transport—passengers	8
Pipeline transportation	22
Radio and television broadcasting	6
Railroads	
Machinery and equipment	14
Structures and improvements	30
Grading and right-of-way improvements	*
Wharves and docks	20
Power plant and equipment	
Hydraulic generating equipment	50
Nuclear generating equipment	20
Steam-generating equipment	28
Steam and other power plant equipment	28
Telephone and telegraph communications	*
Water transportation	20
Water utilities	50

*To be determined according to the facts and circumstances.

By studying the retirement history of a particular type of property, a frequency distribution of property retirement can be obtained. From this frequency distribution a cumulative frequency distribution curve—or survivor curve—can also be plotted as can a probable life curve indicating the potential future life of an asset of any given age. Figure 20 shows a generalized survivor curve, with the corresponding frequency distribution and probable life curves. The frequency curve shows the percentage of the total number of units retired during each time interval. The survivor curve shows the percentage of the total number of units of a given age still in service. The probable life curve gives the total life expectancy of an asset of any given age. The horizontal distance between survivor and probable life curves represents the remaining life of an asset of any age.

Survivor curves are most useful when dealing with "mass" properties, where large numbers of items of relatively low unit value are involved. Typical examples of mass properties include gas, electric, and water meters, distribution mains, utility poles, and similar items.

Just as in analyzing human life expectancy for insurance purposes, analyzing mass properties for depreciation purposes can have value in planning.

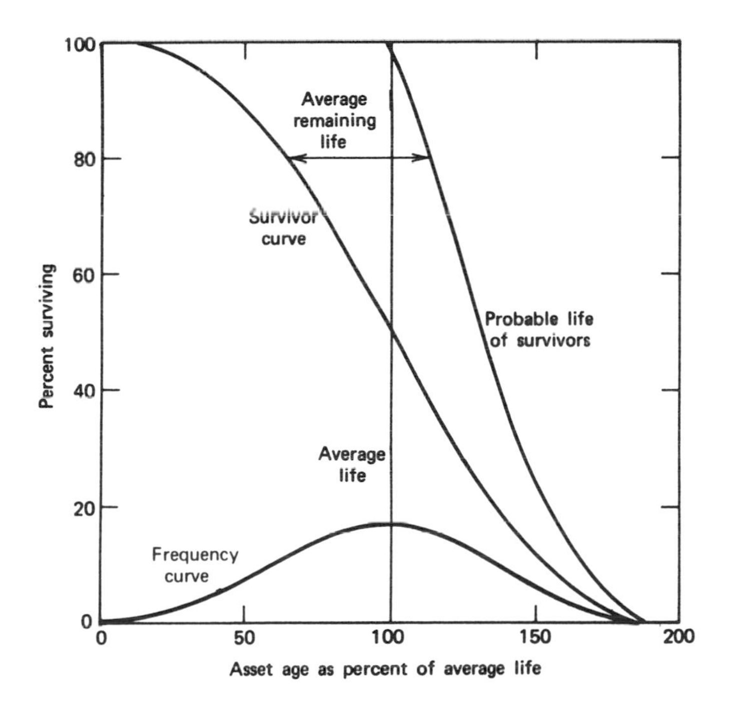

Fig. 20. *Survivor and probable life curve for mass properties.*

The probable future life of an asset can be statistically predicted from its present age, thus providing a useful guideline in deciding at what age it should be retired or replaced—as with humans, whenever the present value of its future services is less than its current replacement cost.

CALCULATING DEPRECIATION CHARGES

Once the economic service life of an asset has been established, either by estimate, analysis, or IRS guideline, there are six different commonly accepted approaches to calculation of annual depreciation charges:

1. The production method.
2. The inventory method.
3. The straight-line method.
4. The sum of years digits method.
5. The declining balance method.
6. The sinking fund method.

THE PRODUCTION METHOD. In the production method, the asset's original cost is prorated on a per-unit-of-output basis. For example, an automobile might be depreciated over 60,000 miles of use rather than on a time basis. The proportion of the initial cost charged during a year, then, is calculated by dividing the number of miles driven that year by 60,000. The underlying assumption in this method is that use alone—and not obsolescence, time, or other factors—causes depreciation. Heavy construction equipment is often handled in this manner, usually on an hours-of-use basis.

THE INVENTORY METHOD. The inventory method bases depreciation on an actual appraisal or count of the property. As such, it is particularly appropriate for property accounts involving large numbers of small items, especially where breakage, loss, or theft are apt to be more important than usage or time. Small tools are sometimes charged on this basis.

THE STRAIGHT-LINE METHOD. The straight-line method assumes a constant amount of depreciation for each year of the asset's life and is found by dividing the original cost (less anticipated salvage value, if any) by the estimated service life. It offers the advantage of simplicity—an important virtue in preliminary economic studies—but may lead to understating investment profitability when income tax considerations are brought in.

THE SUM OF YEARS DIGITS METHOD. This method offers high depreciation charges, along with correspondingly low income tax liabilities during the asset's early years, and is therefore popular. The amount charged in any one year is found by first adding together the digits representing each year in the asset's life; a 5-year life, for example, gives a sum of 15 (1 + 2 + 3 + 4 + 5). Then a proportion of this total is taken for each year. The charge for the first year is 5/15 of the original cost; the second year, 4/15; the third year, 3/15; the fourth year, 2/15; and the last year, the remaining 1/15.

THE DECLINING BALANCE METHOD. The declining balance method, similar to the sum of years digits method, offers accelerated depreciation charges during the early years. Depreciation is taken as a constant percentage of the declining account balance, with the rate usually calculated at twice the corresponding straight-line rate (referred to as the double declining balance method). An asset with a 10-year life, then, is depreciated at 20 percent of its undepreciated balance each year. Since the asset is never 100 percent depreciated this way, the undepreciated portion of the account balance is written off the last year.

THE SINKING FUND METHOD. This is a somewhat antiquated method which assumes that a constant annual amount will be reinvested at some compound interest rate, so that the total principal and interest will accumulate at the end of the asset's life to an amount equal to the asset's original cost. This unrealistic method results in lower depreciation charges during the early years than during the later years and is seldom used except by a few public utilities who wish to maintain a high book value as their rate base.

COMBINATION OF DIFFERENT METHODS. Many companies favor—usually for tax reasons—a combination of a double declining balance method over the first half of an asset's life, switching to a straight-line writeoff for the last half. This approach offers a rapid write-off initially, then later stabilizes the deduction. The halfway point is where the straight-line deduction exceeds the amount deductible by continuing the double declining balance rate.

DEPRECIATION'S EFFECT ON PROFITABILITY

The effect of the choice of a depreciation method on a project's apparent feasibility is illustrated in Tables 28–31. Table 28 shows a project's cash flow using straight-line depreciation, while Table 29 shows the corresponding cash flows when the sum of years digits method is employed. The dollar amounts are all the same in terms of total cash flows over the 5-year period; only the timing of cash flow differs.

When the time value of money is considered, however, the profitability associated with these two cases differs as shown in Tables 30 and 31. By using a DCF approach to determine the profitability of the alternative depreciation

Table 28 Cash Flow Using Straight-Line Depreciation

Year	Cash Outlay	Net Income before Taxes, Depreciation	Depreciation Charges	Net Taxable Income	Income Taxes	Net Income after Taxes	Net Cash Flow
0	150,000	0	0	0	0	0	(150,000)
1	0	50,000	30,000	20,000	10,000	10,000	40,000
2	0	50,000	30,000	20,000	10,000	10,000	40,000
3	0	50,000	30,000	20,000	10,000	10,000	40,000
4	0	50,000	30,000	20,000	10,000	10,000	40,000
5	0	50,000	30,000	20,000	10,000	10,000	40,000
Total	150,000	250,000	150,000	100,000	50,000	50,000	50,000

Table 29 Cash Flow Using Sum of Years Digits Depreciation

Year	Cash Outlay	Net Income before Taxes, Depreciation	Depreciation Charges	Net Taxable Income	Income Taxes	Net Income after Taxes	Net Cash Flow
0	150,000	0	0	0	0	0	(150,000)
1	0	50,000	50,000	0	0	0	50,000
2	0	50,000	40,000	10,000	5,000	5,000	45,000
3	0	50,000	30,000	20,000	10,000	10,000	40,000
4	0	50,000	20,000	30,000	15,000	15,000	35,000
5	0	50,000	10,000	40,000	20,000	20,000	30,000
Total	150,000	250,000	150,000	100,000	50,000	50,000	50,000

Table 30 Profitability Using Straight-Line Depreciation

Year	Net Cash Flow	10.4% Discount Factor	Present Value of Net Cash Flow
0	(150,000)	1.000	(150,000)
1	40,000	0.905	36,200
2	40,000	0.821	32,800
3	40,000	0.743	29,700
4	40,000	0.674	26,900
5	40,000	0.610	24,400
Total	50,000		0

Table 31 Profitability Using Sum of Years Digits Depreciation

Year	Net Cash Flow	11.5% Discount Factor	Present Value of Net Cash Flow
0	(150,000)	1.000	(150,000)
1	50,000	0.897	44,900
2	45,000	0.804	36,200
3	40,000	0.722	28,900
4	35,000	0.647	22,600
5	30,000	0.580	17,400
Total	50,000		0

plans, the straight-line method is found to yield a 10.4-percent return while the sum of years digits method results in a return on investment of 11.5 percent.

Although the absolute difference in apparent yields is not great in this case—11.5 percent against 10.4 percent—this represents a 10-percent variance *on the same investment proposal.* If the client's cutoff point were set at 11.0 percent, the same project that would be approved using one approach to depreciation would be rejected were the other depreciation method used in the investment analysis.

With even the possibility of such an impact, the engineer responsible for the feasibility study must be certain that the approach he takes to depreciation recognizes his client's or his company's best interests—both in terms of future cash requirements and present tax savings.

CHOOSING THE APPROPRIATE METHOD

Although all the different approaches to depreciation accounting eventually recover the original cost of the asset, the choice of a depreciation policy can have an important effect on a project's financial status. The significance of the depreciation decision is not so much in the total amounts involved but rather in the timing of current and future tax-free cash flows. Table 32 shows the annual deductions and undepreciated balances for a $10,000 asset having a 10-year depreciable life, using four different methods of depreciation.

As indicated in Table 32 and illustrated graphically in Figure 21, the accelerated methods result in substantially higher deductions during the early years of an asset's life, and in substantially lower undepreciated account balances during the asset's life.

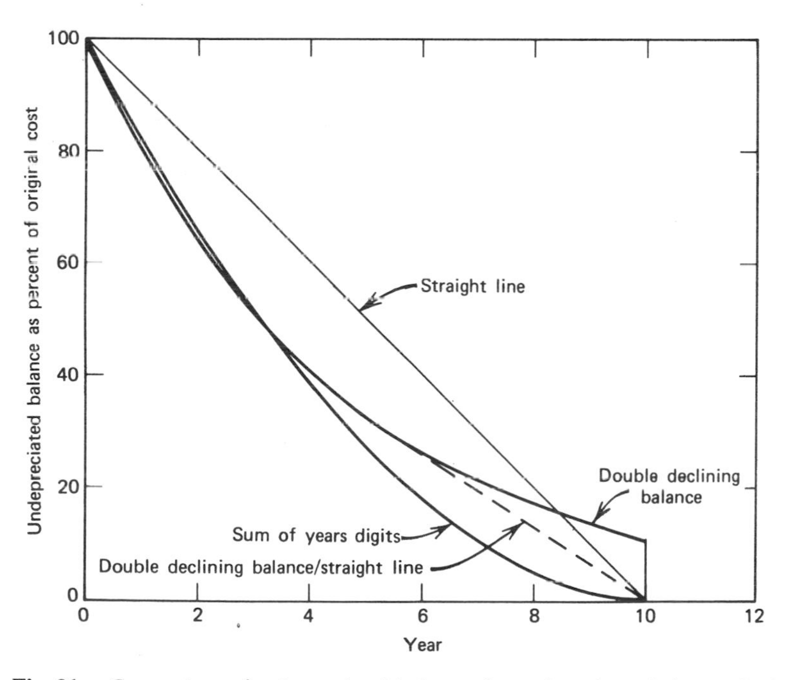

Fig. 21. *Comparison of underpreciated balances for various depreciation methods.*

Table 32 Comparison of Commonly Used Depreciation Methods

	Straight Line		Sum of Years Digits		Double Declining Balance		Double Declining Balance	
Year	End-of-Year Undepreciated Balance	Annual Depreciation Charge	End-of-Year Undepreciated Balance	Annual Depreciation Charge	End-of-Year Undepreciated Balance	Annual Depreciation Charge	End-of-Year Undepreciated Balance	Annual Depreciation Charge
0	10,000	0	10,000	0	10,000	0	10,000	0
1	9,000	1,000	8,182	1,818	8,000	2,000	8,000	2,000
2	8,000	1,000	6,546	1,636	6,400	1,600	6,400	1,600
3	7,000	1,000	5,091	1,455	5,120	1,280	5,120	1,280
4	6,000	1,000	3,818	1,273	4,096	1,024	4,0961	1,024
5	5,000	1,000	2,727	1,091	3,277	819	3,277	819
6	4,000	1,000	1,818	909	2,622	655	2,622	655
7	3,000	1,000	1,091	727	2,098	524	1,967	655
8	2,000	1,000	545	546	1,678	420	1,312	655
9	1,000	1,000	181	364	1,342	336	657	655
10	0	1,000	0	181	1,074	268	0	657
Total		10,000		10,000		8,926		10,000

The straight-line method leaves the highest undepreciated balance throughout most of the asset's life, except for the declining balance method's undepreciated balance near the end of the 10-year period. Switching from the double declining balance to the straight-line method at midlife, though, eliminates this problem handily.

The sum of years digits method gives deductions comparable to the double-declining balance over the first few years and gives larger deductions until the asset's life is about three-fourths over.

The advantage of the higher early-year deductions is apparent when the present values of the annual depreciation charges are compared; here the superiority of the sum of years digits method becomes evident. Figure 22 shows the present values of the 10-year depreciation charges for different interest rates for each of the four depreciation methods.

The ultimate selection of a depreciation method depends on the company's overall financial picture. But it must be recognized that, regardless of interest rates or company policies, depreciation recovers only the book cost —not the total cost—of a capital asset. In computing the true cost of capital, interest must be considered. This is the subject of the next chapter.

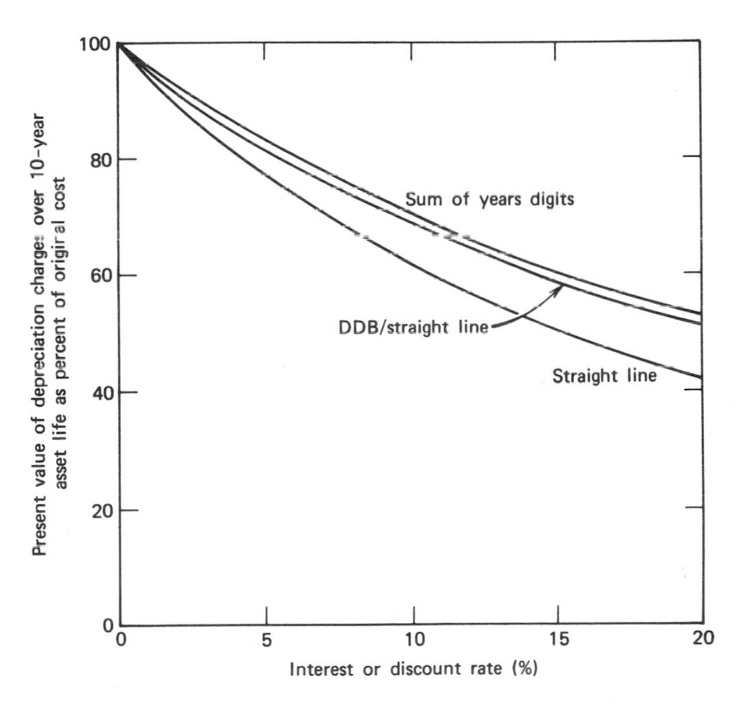

Fig. 22. *Present value of annual depreciation charges for various depreciation methods. DDB, Double-declining balance.*

SUMMARY

Depreciation plays an important role in economic evaluations; it provides the means by which the out-of-pocket cost or book value of a capital asset can be recovered (in an accounting sense) over the asset's economic life. The economic life of an existing asset, or assets similar to those for which experience data have been accumulated, can be calculated by employing actuarial techniques. In new ventures either past experience or IRS guidelines can provide a reasonably sound basis for estimates of asset life. Once the asset life has been established or estimated, depreciation charges can be computed by several different methods, the most common of which are: (1) the straight-line method; (2) the sum of years digits method; (3) the double declining balance method; and (4) a combination of the double declining balance and straight-line methods. The straight-line method is especially useful in preliminary studies, simplicity being its main virtue; it tends to understate the profitability of a new venture and can therefore be considered a conservative approach. Either the sum of years digits or the double declining balance–straight-line combination method should generally be used in a detailed evaluation, depending on the company's financial practices and policies. These methods give higher deductions during an asset's (or project's) early years, thus increasing early-year cash flows and improving the ROI.

7

The Cost of Capital

The cost of capital is an essential part of nearly every economic evaluation. It is particularly important in providing a valid test of project desirability and in helping to select the best available use for a firm's money. The ROI must obviously be at least equal to the cost of capital used in financing the new venture.

Capital can be defined simply as the funds used in the operation of a business. Most firms have available for investment two types of capital: debt capital and equity capital. Debt capital is obtained by borrowing, while equity capital is supplied by the firm's owners, or stockholders, and represents the amount of their ownership in the company. Equity capital also includes retained earnings.

Those who supply the firm's capital—whether lenders or owners—naturally expect a return on their money. Lenders receive their return in the form of interest. Shareholders receive theirs as a combination of dividends and, hopefully, capital appreciation.

The cost of capital, then, has different meanings to different people. To the lender it means the interest he receives on the money he lends; to the stockholder it represents his income from dividends and capital appreciation; and to the company the cost of capital represents the amount that must be earned to satisfy the expectations of the lenders and shareholders who furnished its funds.

From the company's standpoint all that is necessary in determining its cost of capital is to measure the cost of each type of capital employed in the business and to weight those costs according to their relative importance. The resultant figure represents the company's overall cost of money. This overall cost of capital is an appropriate figure to use as a basis for comparison in most economic feasibility studies. Should outside sources of capital be used for financing a specific venture, then the amount paid for only that capital can be used in assessing the proposed venture's merits. But even so, new projects that are expected to earn less than the company's overall cost of capital are, at best, of questionable merit.

Just how the cost of capital is used in a specific study depends to a large extent upon the financial characteristics and point of view of the investing firm. A relatively small firm with a single project under consideration is primarily concerned with the immediate sources of capital for that particular project. In this case, if the project is expected to yield a return in excess of the cost of borrowed money for this specific venture, then the investment is apt to be viewed favorably.

The large corporation is faced with a more complex problem. Often the financing of a specific venture from specific sources also has a secondary effect on the capability of the corporation to raise capital for other projects. For this reason it is often better to consider the problems of raising capital and the problems of determining the expected return of a specific venture as completely independent and distinct problems.

SOURCES OF FUNDS

The cost of capital to be used by a company in financing a new venture depends largely upon the method by which the project is to be financed and the source of the funds used. It is therefore important to consider all the financial details of a project before passing judgment on its desirability or feasibility.

In obtaining the necessary funds for a major project, the company must carefully weigh a variety of outside sources available to it in the financial market.

The characteristics of the various segments of the capital and money markets to which a corporation, or in some cases a government body, can turn to obtain its funds, were discussed in Chapter 2. The major financial markets available to corporations include corporate bonds, corporate equities, commercial bank loans, and commercial paper.

Corporate bonds, equities, and bank loans constitute the most important sources of long-term capital used in financing major new ventures, while bank loans and commercial paper are used extensively as sources of relatively short-term working capital.

THE COST OF DEBT CAPITAL

The cost of short-term borrowings on the money market can be easily determined by reference to the current rates on commercial loans, treasury bills, and other short-term obligations. Bonds, though, present a more difficult problem.

Long-term bonds are bought primarily for their annual interest returns, while short-term bonds may be purchased either for their capital appreciation potential or used simply as a depository for idle funds at a reasonable interest rate until a better opportunity comes along. A bond's overall value as an investment, or its cost as a source of funds, includes both of these elements measured in terms of a composite annual percentage ROI over the remaining life of the bond.

The yield to maturity (YTM) of a bond (its annual interest cost) is determined by four factors:

1. Current market price.
2. Redemption or maturity value.
3. Coupon rate or annual interest payment.
4. Years to maturity.

Market price is the only variable among these four factors, since the annual interest payments are defined by a bond's coupon rate, and the maturity date and value are specified. Consequently, the bond's price is the only factor that can be varied in the marketplace to reflect changing interest rates and money costs.

Given price, coupon rate, and maturity date, the YTM of a bond can be easily computed by conventional investment evaluation techniques. Two approaches can be taken in establishing the attractiveness of a bond as an investment: (1) the present value approach, and (2) the DCF approach.

PRESENT VALUE OF A BOND. The present value of a bond consists of the present value of its maturity value plus the present value of its annual interest payments. The equation for a bond's present value is:

$$P = \frac{M}{(1+R)^N} + C\left[\frac{(1+R)^N - 1}{R(1+R)^N}\right]$$

where P is the present value; M is the redemption or maturity value; C is the annual interest payment determined by the coupon rate; R is an appropriate discount rate, reflecting the short- and long-term money markets; and N is the number of years until the bond matures.

By using this equation a $1000 bond maturing in 16 years and bearing a 5-percent coupon (yielding $50 annually) is, at a 7-percent discount rate, worth

$$P = \frac{1000}{(1.07)^{16}} + 50\left[\frac{(1.07)^{16} - 1}{0.07(1.07)^{16}}\right]$$

$$= 339 + 472$$

$$= \$811$$

The present value of this bond's redemption value is $339, while the annual interest payments are worth $472 at the 7-percent discount rate. Under these conditions, if the bond is sold for less than $811, it will yield (or cost its issuer) more than 7 percent annually. Similarly, if the bond is priced above $811, it will yield or cost less than 7 percent annually over its remaining life.

YIELD TO MATURITY. The annual percentage yield of a bond can best be computed by employing the DCF technique. This approach does away with the necessity of artificially establishing a discount rate and provides a single index for comparing bonds of similar quality but having different prices, coupon rates, and maturity dates. This is the approach used in developing the values published in the bond yield tables used by bond brokers and investors.

The DCF technique is a special case of the present value approach described in the preceding section. Instead of selecting an arbitrary discount rate and computing the bond's present value, the DCF method selects the bond's present value and computes the discount rate that satisfies the prescribed conditions. The bond's present value is assumed to be its current market price, and the computed discount rate becomes, by definition, the annual percentage yield or annual interest cost. Thus the equation is the same as before; only the unknown is different.

By using the example of a $1000 bond maturing in 16 years, carrying a 5-percent coupon, and priced at $740, the equation becomes

$$740 = \frac{1000}{(1 + R)^{16}} + 50\left[\frac{(1 + R)^{16} - 1}{R(1 + R)^{16}}\right]$$

where R represents the annual percentage ROI when the bond is held to maturity. Solving for R in an equation such as this is, as usual, a simple but laborious task, generally accomplished by trial and error. In this case R turns out to be about 7.9 percent:

$$\frac{1000}{(1.079)^{16}} + 50\left[\frac{(1.079)^{16} - 1}{0.079(1.079)^{16}}\right]$$
$$= 296 + 445$$
$$= \$741$$

which is close enough to the $740 market price. The cost of the issuing corporation's long-term debt capital here, then, is 7.9 percent.

If one lacks the patience to go through the calculations associated with the DCF approach, the YTM of a bond can be estimated by using the approximation:

$$\text{YTM} = \frac{2C}{P + M} + \left(\frac{M}{P}\right)^{1/N} - 1.00$$

where, as before, C is the annual coupon rate or interest payment per $1000 maturity value, M the maturity value, N the number of years until maturity; and P the current market price. In using the same example the approximate YTM is

$$\text{YTM} = \frac{2 \times 50}{740 + 1000} + \left(\frac{1000}{740}\right)^{1/16} - 1.00$$
$$= 0.057 + 1.019 - 1.00$$
$$= 0.076, \text{ or } 7.6 \text{ percent}$$

as compared with the actual 7.9 percent determined previously. If set of bond yield tables are not available, this approximate method is quite useful in obtaining a quick and easy measure of a bond's investment value or interest cost, or in establishing a trial rate to use in subsequent DCF calculations.

THE COST OF EQUITY CAPITAL

The cost of equity capital must be recognized as a real economic cost of doing business, even though it represents an "opportunity cost" to investors rather than a contractual obligation on the company's part. It is in fact an obligation to earn, rather than to pay, a given amount. If a publicly held company is to maintain its ability to attract equity capital, the company must earn for its investors at least as much as they could earn by investing their money elsewhere at comparable risk. Should the company cease to be attractive to equity investors, it is likely to become equally unattractive to other sources of capital.

Estimating a realistic cost for equity capital requires considerable judgment. The price of a company's common stock reflects the investing public's appraisal of the present worth of the company's future earnings. The investor is in effect purchasing an annuity made up of a stream of future dividends or earnings, plus an opportunity to sell his shares whenever he wishes at the prevailing market price.

The cost of equity capital, then, includes two elements: dividends and the possibility of capital appreciation; both in turn are related to earnings. Just how to combine these two elements into a single figure representing the cost of equity capital has always been a perplexing problem.

There are many ways to approach this problem, but only two of the simplest ones are presented here. Both of them are commonly used.

One approach uses simply the ratio of present earnings to present market price, on the assumption that purchase of a share of equity stock at a given price will return to the investor a certain amount per year in earnings. Or, from the company's standpoint, a certain amount of present earnings must be given up to attract each dollar of present equity capital. This ratio must therefore represent the cost of equity capital to the company.

Another commonly employed method of estimating the cost of equity capital is from the investor's standpoint rather than the company's. First, the prospective purchaser of a share of capital stock expects to receive a definite cash return on his investment—the dividend. He probably also expects the stock's value to increase with time, as the company's retained earnings are reinvested in the business to generate more earnings.

Simply adding the dividend yield to the anticipated earnings growth rate, then, gives a measure of the investor's expectations. For example, if the company pays a 5-percent dividend and earnings are expected to increase at a rate of 5 percent annually, then 10 percent could be said to represent the amount that the company must earn to satisfy the investor and therefore is its cost of capital. This method is probably as justifiable as any other.

A low earnings/price ratio (or high price/earnings ratio) is generally associated with a high anticipated growth rate, and low dividend yields are characteristic of fast-growing companies. Consequently, equity capital may be cheaper for newer, fast-growing firms, while debt capital is usually cheaper for mature companies with high dividend rates.

THE OVERALL COST OF CAPITAL

After a company's capital structure and the cost of each type of capital used in the business have been determined, the overall cost of capital can be easily calculated.

The financial structure of major companies in different fields varies widely, as shown in the following table.

	Capital Structure (%)			
Industry	Debt Capital	Equity Capital	Other	Total
Oil	17.6	79.0	3.4	100.0
Steel	27.2	67.2	5.6	100.0
Utility	44.2	52.6	3.2	100.0
Chemicals	5.2	82.9	11.9	100.0
Manufacturing	2.7	93.0	4.3	100.0

The cost of capital to these large corporations can be expected to vary equally widely, as do their preferred methods of financing.

Consider, for example, a company having the following financial structure.

	Percent of Total Capital	Cost (%)	Weighted Cost (%)
Short-term debt	5.0	6.0	0.30
Long-term debt	15.0	8.0	1.20
Shareholders' equity	80.0	10.0	8.00
Total	100.0		9.50

The company's short- and long-term debt costs can be easily obtained from published financial statements. Shareholders' equity includes all capital stock—common and preferred— and also retained earnings and other long-term reserves.

The cost of conventional debt financing is the average market rate expected on new borrowings and is shown in the example at 6.0 percent for short-term and 8.0 percent for long-term obligations; the exact figures can be established from the current market value of the firm's outstanding notes and bonds and may be well over 10 percent, depending on the prevailing money and capital markets. Equity capital is assumed to be priced at about 10.0 percent in the example.

As shown in the example, weighting each type of capital by its cost results in an overall figure of 9.50 percent.

Should the company's capital structure change, or should the interest rates or other factors relating to the cost of a particular source of capital vary, then the company's overall cost of capital would change accordingly.

If in the preceding example the company's financial structure were 10.0 percent short-term debt, 30.0 percent long-term debt, and 60.0 percent equity capital, its overall cost of capital would drop to 9.0 percent. This company obviously would prefer debt to equity financing.

In the present value approach to measuring profitability, the cost of capital serves as the appropriate discount rate for future earnings; in DCF analysis it might represent the cutoff point beneath which investment proposals would be rejected. But regardless of the method used in appraising new investment opportunities, a realistic measure of investment desirability cannot be obtained without at least considering the company's overall cost of capital.

SUMMARY

Capital represents the funds used in the operation of a business, and its cost provides a useful guideline in evaluating project desirability and in selecting appropriate uses for available money. A company's cost of capital depends on its capital structure and its methods of financing and sources of funds, as well as on the general economic conditions reflected in the capital and money markets. Capital can be obtained either by borrowing (debt capital) or by selling ownership (equity capital). Equity capital is especially important for new or fast-growing companies, while debt capital is the less costly and generally preferred source of outside funds for well-established firms. The expected return on new projects should be higher than the cost of capital used in directly financing the project; if internal funds are employed, then the proposed project should earn at a rate higher than the company's overall cost of capital. Projects expected to earn at or less than the company's cost of capital should be avoided whenever possible.

8

Cost Analysis

Every business sells either products or services and in its operation must incur certain costs. The analysis of these costs—including their identification, measurement, allocation, and control—is an extremely important activity in every business. Regardless of where the cost analysis is being performed, its purpose is essentially the same: to provide factually accurate and objective information useful in pursuit of the firm's objectives. The specific purpose of a cost analysis may be to provide information to be used in the design of a rate schedule in the case of a public utility, in the development of an appropriate and reasonable pricing policy for a manufacturer, or as a guide for establishing rental or leasing terms in a real estate venture. It may be used as a basis for preparing engineering economic studies, as a planning tool in sales promotion activities, or as an aid in cost control and capital budgeting.

ELEMENTS OF COST

In an accounting sense cost refers primarily to the expense items on the firm's operating statement such as property taxes, interest, and depreciation, in addition to materials, operating labor, maintenance, supplies, and other conventional expense items. However, the accountant may exclude from his definition of cost the return on equity investment (or profit) and income taxes associated with this return.

However, the engineering approach to cost includes, in addition to capital and operating expenses, an appropriate return or profit on the capital invested in the business. This approach is necessary in evaluating the economics of a project. Cost, to be completely realistic, must refer to the *entire* economic activity; total cost, then, represents the firm's revenue requirement, the total amount to be received from sales of its goods or services. Cost analysis shows where the money comes from, where it goes, and why.

Only through this engineering concept of cost can the reasonableness of a firm's rates or prices be appraised, as measured by the criterion of whether or not a fair return or profit has resulted after all capital and operating charges are deducted from total revenues.

In serving its customers or clients, a firm must perform a variety of different operations, each of which fulfills a specific function and contributes a specific cost. Functionally different operations may be performed at the same physical location, and similar functional operations may be performed at different physical locations. In making a cost analysis, primary emphasis should be on the functional aspects; costs should therefore be grouped by functional rather than physical locations.

TYPES OF COSTS

Involved cost structures and accounting systems sometimes obscure the relationships between sales, costs, and profits. To help clarify these relationships, costs are usually broken down into either two or three broad categories: fixed, variable, and, sometimes, semivariable.

FIXED COSTS. Fixed or overhead costs are costs whose magnitude depends upon the capacity of a system rather than upon the conduct of an activity within the system. These fixed charges are largely a function of time and are the same during any designated time period regardless of the level of output. They may be prescribed by contract (such as office rent), or incurred just to insure the existence of an operating organization (such as accounting and personnel departments); or they may consist of a utility's capacity costs associated with investments in its plants or distribution lines.

VARIABLE COSTS. Expenses that move in close proportion to changes in production—which can be attributed to a specific function and are therefore chargeable in their entirety against that particular function—are known as variable or direct costs. An electric generating plant's fuel cost, for example, varies directly with the net generation. Similarly, a gas distribution utility might refer to its "commodity" costs—the cost of natural gas purchased from the supplier's pipeline—as a variable cost. Some firms may have relatively few costs that can be considered completely variable, while other types of businesses may find a large proportion of their total costs to be of a variable nature.

SEMIVARIABLE COSTS. In many operations costs can be categorized as either fixed or variable. Some costs, however, may not fit well into either of these categories but assume some of the characteristics of each. These must be classified as semivariable. Semivariable costs are often associated with a firm's need to keep a portion of its physical facilities and personnel intact almost regardless of the level of business. Some companies, for example, must keep a nucleus of competent professional personnel on the payroll at all times to maintain a capability to do business. Unutilized professional staff time assumes the characteristics of an overhead or fixed cost, while other staff salaries are variable costs. A utility's customer costs might also be considered semivariable costs, consisting of accounting, collecting, and sales promotion expenses.

THE COST ANALYSIS PROCEDURE

The cost analysis procedure involves four separate and distinct phases:

1. Break down total costs by functional locations.
2. Classify cost elements according to their causes.
3. Determine cost responsibilities.
4. Allocate costs according to the responsibilities involved.

COST BREAKDOWN. The initial cost breakdown is by function. All costs are presumably incurred for some specific purpose; the functional breakdown of costs simply defines the purpose for which each cost is incurred.

COST CAUSATIONS. All costs can be attributed to a firm's capacity to do business, to the amount of business it actually does, or to a combination of these two elements. All costs, then, can be classified as fixed, variable, or semivariable in nature.

COST RESPONSIBILITIES. In establishing cost responsibilities it is necessary to define exactly what resources are employed in serving each class of customer or in providing each type of service, and to identify the specific costs associated with each of these resources.

COST ALLOCATION. After all elements of cost accruing at different functional locations have been classified according to their causes and cost responsibilities have been defined, total costs can be allocated to the various products or services in accordance with their particular cost responsibilities.

INCREMENTAL COST ANALYSIS

Incremental, or marginal, cost refers to the additional out-of-pocket cost incurred in producing each additional unit of output without an additional investment in production facilities. Rather than employ conventional accounting techniques in estimating and allocating overhead and other indirect loadings, only the actual cash flows are considered in incremental cost analysis. Thus alternative projects or courses of action are compared strictly on a cash basis.

In theory, as long as each unit of additional output generates revenues greater than the out-of-pocket cost of obtaining the added output, it pays to increase production. When, finally, marginal cost is *equal* to marginal revenue, the optimum level has been reached.

Consider, for example, the situation described in Table 33. Here the alternatives range from producing zero to 12 units of additional output without affecting the level of fixed costs, which remain at $150 throughout this range; total revenues, meanwhile, are assumed to be a constant $150 per unit of output. Direct production costs can be expected to decrease slightly on a per-unit basis as the production rate increases, while other variable unit costs will probably increase. These other variable costs might represent additional promotional and advertising expenses, distribution costs, or price discounts required to dispose of the additional production; they give the same effect as a price reduction, also commonly associated with increased production.

As indicated in Table 33 and illustrated graphically in Figures 23 and 24, maximum profits will occur at a production level of between eight and nine units. This holds true whether or not fixed costs are included in the analysis, since only the *marginal* cost–*marginal* revenue relationship determines the optimum production rate.

This incremental cost concept plays an important role in the economic theory of cost and production; but as is the case with many important concepts in economic theory, its application to real-life problems is not always clear.

When applying the incremental cost concept, it is frequently necessary to stand back and take an overall view of the project being studied and its relationship to the company's entire operation. Incremental improvements in profits for a given project may be the best obtainable for that project but

Table 33 Example of Incremental Cost Analysis

Units of Production	Fixed Costs	Direct Production Costs	Other Variable Costs	Total Cost	Total Revenue	Total Profit	Marginal Cost	Marginal Revenue
0	150	0	0	150	0	(150)	0	0
1	150	100	20	270	150	(120)	120	150
2	150	195	30	375	300	(75)	105	150
3	150	285	40	475	450	(25)	100	150
4	150	370	55	575	600	25	100	150
5	150	450	80	680	750	70	105	150
6	150	525	110	785	900	115	105	150
7	150	595	160	905	1050	145	120	150
8	150	660	225	1035	1200	165	130	150
9	150	720	320	1190	1350	160	155	150
10	150	775	450	1375	1500	125	185	150
11	150	825	640	1615	1650	35	240	150
12	150	870	900	1920	1800	(120)	305	150

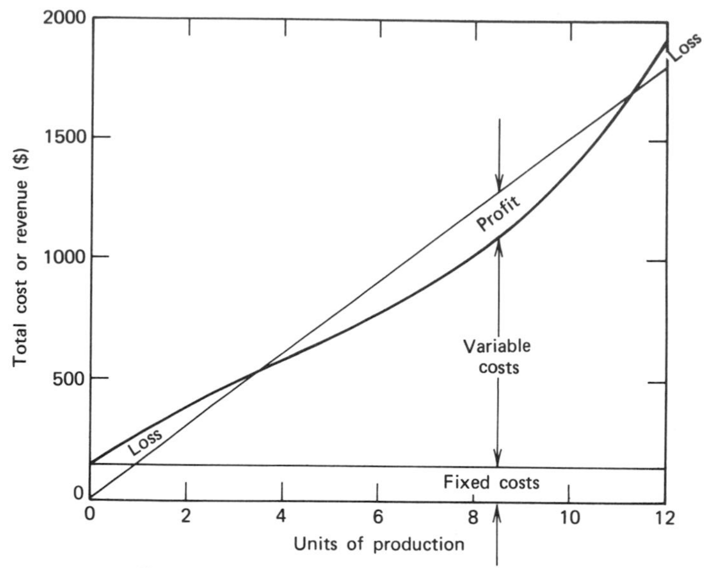

Fig. 23. *Example of incremental cost analysis.*

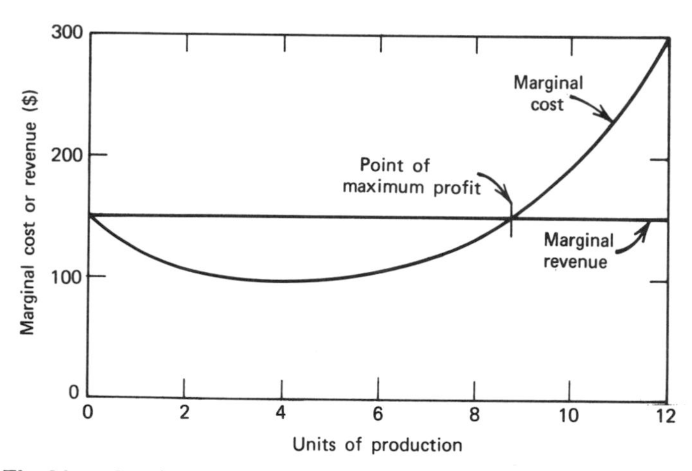

Fig. 24. *Identifying optimum production level by incremental cost analysis.*

still fail to meet the company's profitability objectives or standards—in which case the project should be deferred in favor of others that might generate a higher return.

Since incremental cost studies are necessarily tied to past decisions, the time will come when the incremental project had best be abandoned entirely, even if an additional investment would in itself produce a satisfactory return.

Whenever the project as a whole—even including the considered incremental investment—is failing to meet the company's goals, then that project is exerting a negative incremental effect on the company's total operation. The possibility of dropping the entire project should then be considered.

The incremental cost concept, however, illustrates a point that is extremely important in most engineering economy studies; economic evaluations should be based on the anticipated effects of the best *presently* available alternative and not tied to bad decisions made in the past. The only costs that need be considered are those over which some degree of control can be exercised by present actions.

ANALYSIS OF CASH AND CAPITAL BUDGETS

The fact that business operations are complex and difficult to forecast is sometimes used as an excuse for not operating a firm under close budgetary control. Complexity of operations, however, indicates the importance of emphasizing, rather than ignoring, the potential benefits of a thoughtfully prepared analysis of budgetary controls.

A budget, by definition, is a "statement of expected revenues and expenditures." Budgetary control, then, refers to an overall plan for analyzing and monitoring these expected revenues and expenditures by setting standards against which actual performance can be measured. When properly designed, the budget program can detect inefficiencies, fix responsibilities of individuals involved, and help insure the best possible utilization of the firm's physical and financial resources.

The basic purpose of any budget is to provide top management with the information needed to make intelligent, rational decisions; a budget's effectiveness can be measured only in terms of the usefulness of the information it provides.

One of the primary aims of a budgetary system is to aid in planning for future needs by establishing an action plan for management—a plan that will reveal the amount and timing of future resources needed to meet the firm's objectives. In this sense the budget can serve management as an economic model of the firm, which can be used to evaluate proposed changes in objectives or operating procedures.

Another important function of the budget is to provide a means of coordinating different activities within the firm, thus lessening the possibility of conflicts between departments. Each department can see exactly where and how it fits in with the firm's overall programs, what is expected of it, and how it is expected to proceed.

A budgetary program also provides an effective mechanism for controlling costs within the firm by assigning specific individuals the responsibility for controlling designated segments of the budget and providing these individuals with guidelines for handling of funds.

Finally, the budget program offers a framework for providing top management with a broad perspective of the firm's entire operation without requiring excessive involvement in operational nonproblem areas.

The budget ties in closely with the firm's short- and long-range plans. It can in fact serve as a comprehensive financial blueprint of the firm's future.

While any of the firm's resources can be budgeted, top management's primary concern is usually with the financial budget. Financial budgets can be broadly categorized as either capital budgets or cash budgets and may be either short range (1 year or less) or long range (usually from 1 to 10 years). The capital budget is primarily considered a planning tool, showing where the firm hopes to find itself on a given date in terms of a projected balance sheet. The cash budget, however, is an operational tool, indicating expectations rather than plans.

In both cash and capital budget analysis, the amounts should be as precise as possible with the information available at the time of preparation. Naturally, the degree of precision varies inversely with the length of time over which the budget is projected. There should be at least enough detail to recognize variations from the expected values and to permit revisions whenever necessary.

CAPITAL BUDGETS. The capital budget refers to estimated expenditures on capital acccount items—the firm's plans for replacing, improving, or acquiring capital assets. It is concerned with how much, on what, and when the firm's funds will be spent. Projected balance sheets constitute an important part of the capital budgeting process, reflecting anticipated changes in the firm's financial strength and providing a basis for evaluating financial effectiveness. While not directly included in the capital budget, formal procedures for authorizing, analyzing, and reviewing all major capital expenditures should be developed and tied in closely with the budgeting process.

CASH BUDGETS. The cash budget shows the projected flow of funds as related to the firm's operations. Cash budgeting involves the estimation, management, and control of these cash flows, summarizing anticipated receipts and disbursements for a forthcoming period. The cost figures included in the cash budget should reflect reasonably efficient operations and realistic operating conditions. Thus the cash budget serves as an operating aid. In some respects the cash budget is to the firm's financial operations as standard costs are to factory production, in that actual performance is measured against predetermined standards. Again, the standards must be realistic, not perfect; when measured against a standard of perfection, everyone is inefficient.

There are five major steps involved in the analysis or preparation of a firm's budget program. The first four deal with developing the budget itself, while the fifth is concerned with its implementation.

BACKGROUND DATA. The first step in preparing a budget is to accumulate data on the firm's past activities and operations. The information collected should include all pertinent financial data, as well as the number, types, and sizes of projects conducted, clients, locations, personnel utilization, and other operating indicators.

MARKETING ASSUMPTIONS. The second step in preparing the budget is to develop basic assumptions regarding the nature of the market for the firm's products or services as a whole. The market forecast in this step indicates the total volume of business that must support *all* firms.

THE COMPANY PLAN. The next step in budget preparation is to develop basic assumptions regarding the firm itself, including its purposes and objectives, its methods of operation, the special fields to be emphasized, the amount of time and money to be devoted to sales promotion and client contact, new capabilities to be acquired, and the availability of and requirements for personnel, capital, and facilities. This step tells how effectively the firm can compete in the markets defined in the previous step.

MAKING THE PROJECTIONS. The final step in preparation of the budget consists of making estimates of cash and capital flows for the future based on a careful analysis of historical data on past performance, estimates of future operating performance, and assumptions of what the future holds for the business in general. The resulting figures should reflect management's best estimates of future sales, staffing, salary and wage levels, materials costs and other major expense items, along with any plans for expansion of facilities or other major capital commitments. In all cases the individuals responsible for carrying out each phase of the budget should participate in its preparation.

ACTION PROGRAM. After the budget has been prepared and approved, each future decision must be weighed to insure compatibility, or to identify causes of variance, with the projected figures. By continually reviewing all figures and assumptions included in the budget and making appropriate revisions in light of changing conditions, the budget can be made a practical and valuable management tool.

COST-BENEFIT ANALYSIS

The capital available for investment is always limited, and it is important to make sure that whatever funds are available are allocated and spent on a consistent and objective basis. This requires developing some measure of project worth to weigh against estimates of project cost.

This is where cost-benefit and cost effectiveness analysis come in. The cost-versus-benefit approach is becoming increasingly important, especially in federal government projects. Standard procedures for cost-benefit analysis have in fact already been developed for use in water resources projects and defense planning and are equally desirable in connection with recreation, highway, urban renewal, and public health programs.

Cost-benefit analysis offers several distinct advantages which cannot be obtained otherwise:

- It focuses attention on key budget decisions, recognizing both immediate and long-range costs and consequences more clearly.
- It identifies alternative ways of accomplishing the same goals better or more cheaply, or better ways of using the available money for different purposes.
- It helps identify unprofitable or outdated programs or activities and provides the facts needed to make or support a decision to terminate them.
- It enables management to assert tighter control over several independent divisions or departments. This is because there is increased control over both capital outlays and operations.

While cost-benefit analysis does not make decisions for management, it does make the costs and consequences of alternative actions clearer for the decision maker. By presenting the basic information on a proposed project, it practically forces a close examination of alternatives and also serves to eliminate any possible duplications.

The biggest problem in cost-benefit analysis is to measure objectively the benefits accruing from a proposed expenditure. It is often difficult even to foresee the consequences of a project, let alone to measure the values or benefits of these consequences in objective or quantitative terms. Even so, measurable costs can be weighed against measurable benefits; then the *unmeasurable* benefits can be examined to see if they might be impressive enough perhaps to justify the necessary measurable costs.

Social values are the most difficult to quantify. Nevertheless, cost-benefit analysis is especially important for projects in which investments that might not be financially attractive to private business are worthwhile from a social standpoint. These social considerations in fact largely dictate the conditions under which cost-benefit analysis is essential and explain why government-sponsored programs are particularly well suited to this particular type of analysis.

There are three broad types of programs that of necessity must usually be government-supported and that require some assessment of intangible or hard-to-measure values. They deal primarily with peculiar conditions of consumption, production, and timing.

CONSUMPTION. The usual procedure in a free-enterprise system is for the user to pay for the goods or services he consumes, the amount he is willing to pay indicating the value of the commodity to him. But charges for some commodities either are not collectible, or collection is difficult or impractical. For example, it is unwieldly to charge only the direct users for their use of police and fire protection, highways, national defense, outdoor recreation facilities, public health measures, and many municipal services. Since a collective benefit is created by these services, it may be desirable to provide them even though the amounts collectible from the direct users are not enough to cover the services' costs.

PRODUCTION. Economies of scale may demand production facilities so large that private industry cannot possibly provide the necessary resources. Urban redevelopment, large hydroelectric projects, and highway systems are examples falling in this category.

TIMING. In some cases present prices do not adequately reflect the true long-range importance of projects, since the short-term outlook of most private investors precludes, for them, consideration of projects involving only long-term benefits. For example, the present costs of exploiting natural resources do not necessarily reflect the importance of these resources to future generations. Similarly, there is the possibility that long-term effects of urban renewal may ultimately far outweigh the immediate benefits.

The approaches to cost-benefit analysis are basically the same as the approaches used by private investors in evaluating investment proposals, the main difference being only in the means by which benefits are measured. The first step in the cost-benefit study is to project the physical output by years in whatever terms are appropriate—passenger miles or kilowatt hours, for example. Next some estimate must be made of the social value of these physical outputs, in dollars if possible. These two steps provide an estimate of the gross social contribution of the project. After they have been completed, there are two popular methods of comparing costs and benefits.

ANNUAL COST METHOD. After the benefits have been computed for a typical or average year, the comparable operating and maintenance costs are computed for the same year. Then the initial capital costs are converted to an equivalent annual cost basis by amortizing them over the expected life of the project. These annual capital costs, added to the annual operating and maintenance costs, give the total annual project cost. The ratio of the total annual cost to the gross annual benefits, then, gives the cost/benefit ratio to be used in comparisons with similarly derived ratios for other projects.

PRESENT VALUE METHOD. In the second approach the annual operating and maintenance costs for each year are deducted from the gross benefits for the same year, giving the annual net benefit for each year. The resulting figures are then discounted back to the initial year, their total representing the present value of future benefits. The ratio of the initial capital expenditure to this figure is the cost/benefit ratio. This approach can also employ an internal rate-of-return technique similar to the DCF method of investment evaluation.

The choice of method is not critical, since all have the same results if properly used. The important issues are to decide what benefits are to be included, how to value them, and what interest rates to use in amortizing the capital costs or discounting the operating costs involved.

While subjective measures of benefits are satisfactory for comparing alternative ways of accomplishing the same thing, objective measures are usually necessary to compare different programs having different objectives. An early question to be answered is whether or not the benefits *can* be estimated reliably enough to even justify the expenditure for making the analysis.

Dollars naturally constitute the most convenient unit of measurement. One of the best clues in assessing the benefits of any project is always either the cost of providing an equivalent service through alternative means, or the cost of not providing the service at all.

Here are just a few ways that have been used to express benefits in dollars for different types of projects.

PUBLIC HEALTH. The cost of curing a disease serves as a measure of the benefits of preventing its occurrence. This measure is extremely conservative, since the value to an individual of not having some dreaded disease far exceeds the direct cost of this treatment.

URBAN RENEWAL. The difference in property values before and after the elimination of blight and slums is certainly one direct monetary indicator of the benefits that accrue from such programs.

HIGHWAYS. Decreased travel time and travel distance attributable to new highways are two of the many factors that might be translated into monetary terms as a direct measure of highway benefits.

RECREATION. Some indication of the monetary value of public recreation facilities can be obtained by determining accurately what amount people are willing to pay elsewhere for the use of similar recreational opportunities.

TRAFFIC SAFETY. The benefits of accident reduction can be expressed in terms of the monetary value of property and lives saved. Cost-benefit analysis can point out the specific areas where money could best be spent—such as in improved highways, more rigid law enforcement, motor vehicle inspection programs, and the like.

SUMMARY

Cost analysis provides accurate and factual information for management use in working toward a firm's goals and objectives by describing the firm's entire economic activity. Costs may be of a fixed, variable, or semivariable nature, depending on how they are related to the firm's capacity to do business and to its scale of operations. The cost analysis procedure involves four broad steps: (1) attributing costs to functions; (2) classifying costs by causes; (3) determining cost responsibilities; and (4) allocating costs according to responsibilities. Several special fields of cost analysis are also important, including analysis of incremental coasts, cash and capital budgets, and cost-benefit relationships. Incremental cost analysis is a useful tool in identifying optimum production levels or operating scales, while budget analysis serves as a valuable management planning aid. Cost benefit analysis provides a quantitative basis for making decisions in relatively subjective or intangible areas—especially in connection with large government-supported programs conducted for the public good—which would otherwise have to be made almost intuitively.

9

Cost Estimation

Nothing improves the output of an engineering-economic study more than a good input; meaningful conclusions can be drawn only from meaningful data. Cost estimates usually constitute one of the main inputs for economic evaluations.

A variety of estimating methods is available to the engineer, depending on the purpose of his estimates. In general, there is a choice between three levels of detail and accuracy in cost estimating:

1. Order-of-magnitude estimates.
2. Semidetailed or conceptual estimates.
3. Detailed estimates.

While there is no substitute for complete, detailed takeoffs and pricing when a final proposal is submitted for management approval at a guaranteed price (such as in a contractor's fixed-price bid), most feasibility studies require far less sophistication. It is important, though, that the estimator be aware of the limitations of whatever estimating method he employs.

THE COST OF ESTIMATING ACCURACY

The cost estimate can within certain limits be made as accurate as the engineer's company or client is willing to pay for. The more accurate the estimate, the more time required for its preparation and the more design data that must be developed prior to making the estimate.

Figure 24A shows the relative costs associated with the preparation of estimates having a specified level of accuracy, based on about a $1 million project. Estimates on larger projects require a lower expenditure per project dollar, while smaller projects involve higher percentage estimating costs than shown here.

For example, an estimate accurate enough on which to base a fixed-price bid—that is, within ±5 percent—would be expected to be about 4 percent of the total project cost, or $40,000 on a $1 million project.

An estimate of adequate accuracy for budgeting purposes, though—say within ±10 percent—would cost 1.5 percent of the total project amount, or

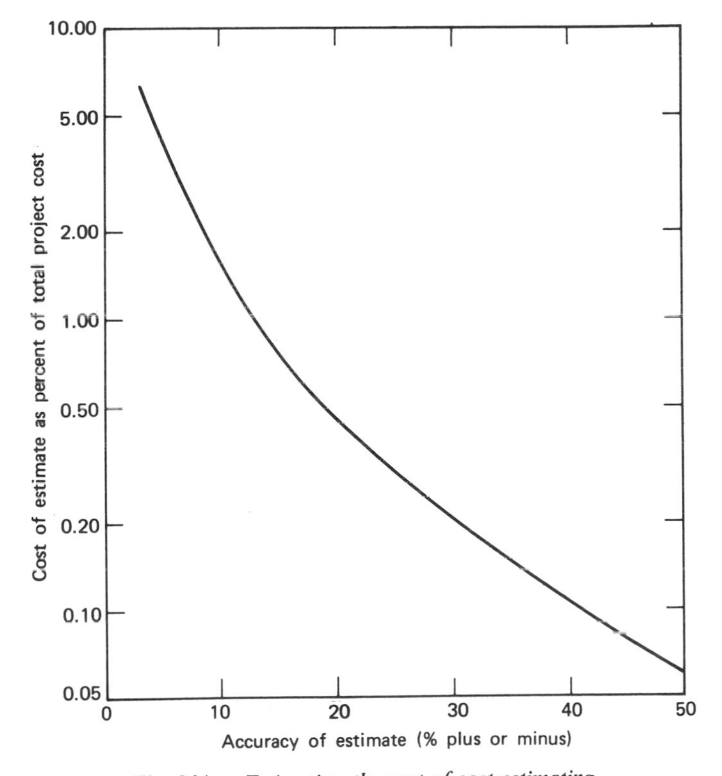

Fig. 24A. *Estimating the cost of cost estimating.*

$15,000 in this case. An order-of-magnitude estimate, accurate only to within ±30 percent, could be made for only 0.2 percent, or about $2000. For $600 a "seat-of-the-pants" or "ballpark" estimate could be had, accurate within ±50 percent.

Ultimately, the actual cost of an estimate depends upon the estimator's skill and the availability and quality of the cost data with which he works.

ORDER-OF-MAGNITUDE ESTIMATES

The order-of-magnitude estimate offers a relatively low level of accuracy, varying by as much as 30–50 percent and sometimes even more from the true project cost, and seldom coming closer than within ±15 percent of the actual cost. To achieve any better accuracy requires that the estimator be thoroughly familiar with the types of work under consideration and that he have access to pertinent cost data on similar projects. The sacrifice in precision, though, is often justified by the ability to screen a large number of alternative projects in a short time.

The value to be gained from short-cut estimates depends largely on the skill, experience, and judgment of the estimator. Estimates must be made with care and applied with caution and full recognition of their limitations. But even though considerable accuracy is sacrificed in the short-cut estimate, it nevertheless can satisfy the need for quickly screening a wide range of alternatives. If warranted, the project in question then can be appraised in greater detail through more precise estimating methods.

Order-of-magnitude estimates can be based on cost capacity relationships, ratios, or physical dimensions.

Cost-Capacity Relationships

One of the most useful short-cut estimating techniques—one that is often used in chemical process industries and by cost engineers in other fields but which is seldom applied to civil engineering projects—employs the cost-capacity relationships among similar projects of different sizes. This method —sometimes called "factoring"— usually can be made to yield acceptable accuracy for study purposes.

The factoring method of short-cut cost estimating involves these steps:

1. Start with a recent project of known cost similar in characteristics (but not necessarily similar in size) to the one being estimated.
2. Convert the cost of this previous project to a current basis using an appropriate index to correct historical costs both for time and location.

Some of the major general-purpose cost indexes used by estimators include the *Engineering News-Record Construction Cost and Building Cost Indexes*, the *Marshall and Stevens Installed Equipment Cost Index*, the *Nelson Refinery Construction Cost Index*; and the *Chemical Engineering Plant Construction Cost Index*. In addition to these, there are numerous special-purpose cost indexes compiled and published by various private firms and governmental agencies such as the Department of Commerce, Department of the Interior, Bureau of Public Roads, and Bureau of Reclamation. With the variety and availability of published indexes covering virtually all types of construction, the estimator should have little difficulty in selecting an index appropriate to his interests.

It is important for the estimator to know exactly what is included in whatever index he uses, however. Important considerations include: frequency of publication; inclusion of labor productivity factors, to reflect improved productivity of field labor due to advances in equipment and technology as well as design innovations in the facilities being constructed; and provisions for geographic differences in wage rates, materials costs, and labor productivity.

3. Define the relative size of the two projects in terms of capacity (gallons per day or installed kilowatts, for example), and end product units (number of classrooms, hospital beds, or apartment units), or physical dimensions (square feet, lineal feet, cubic feet, etc.).

4. Taking the three known quantities—the size and cost of the first project and the size of the second project—solve for the unknown cost of the second project by using an exponential power relationship:

$$C_A \div C_B = (S_A \div S_B)^x$$

where C_A represents the cost of project A, C_B is the cost of project B, S_A is the size of project A, and S_B is the size of project B measured in the same units. Exponential x defines the cost-capacity relationship.

The exponential factor varies according to the type of project being considered but usually is in the 0.6–0.8 range. However, factors as low as 0.2 and as high as 1.0 are not uncommon. Steam-electirc generating plants, for example, usually have a cost-capacity factor of about 0.8. Waste treatment plants employing both primary and secondary treatment usually range between 0.7 and 0.8. Large public housing projects also average about 0.8. But steel storage tanks may have cost-capacity factors as low as 0.4 or as high as 0.8, depending on their shape.

An old estimating rule of thumb in manufacturing states that doubling the production increases costs by half; this represents an exponential factor of 0.585.

To illustrate how this short-cut "factoring" technique works, assume a 100-MW steam generating plant was built in 1960 at a cost of $22 million, with the *ENR Construction Cost Index* at 800. In estimating the present cost of a similar 350-MW plant, with a corresponding cost index of 1000, the procedure is as follows.

Current cost of 100-MW plant:

$$\$22 \text{ million} \times (1000 \div 800) = \$27.5 \text{ million}$$

Current cost of 350-MW plant:

$$27.5 \text{ million} \times (350 \div 100)^{0.8} = \$27.5 \text{ million} \times 2.72$$

The estimated present cost of the 350-MW plant, then, is about $75 million.

The development of appropriate cost-capacity factors for use in making short-cut cost estimates obviously requires the analysis of numerous completed projects. These data fortunately are available in several easily obtainable published sources. When the data are plotted on log-log graph paper, they usually show a straight-line relationship, reflecting this exponential economy of scale. The appropriate cost capacity factors can be calculated after fitting a straight line to the points. Table 33A shows typical cost-capacity factors for a variety of plants and facilities.

Similar to any other short-cut method, the factoring method can result in sizable errors if the estimator is not familiar with the project under consideration or with the inherent limitations of the estimating technique. Particular care must be exercised in selecting an appropriate exponent for factoring. For example, if an exponent of 0.7 were used in estimating the cost of a proposed project having five times the capacity of a previous one when the correct exponent is actually 0.8, an error of about 17 percent in the final estimate would result. If the difference in project size were greater, the error also would be greater; nearly a 26-percent error would be associated with a 10:1 ratio of project sizes.

EQUIPMENT RATIOS, PHYSICAL DIMENSIONS, OR WEIGHT. Some projects involving a few major items of equipment that can be accurately priced can be estimated by applying a multiplier to the cost of these major items. Ratio estimates are often used in chemical process industries in which specialized equipment makes up the major portion of the total project cost.

Physical dimensions or weight can also be used to estimate the approximate cost of many different types of projects. Building cost estimates, for example, can be made in terms of square feet of floor space or cubic feet of building volume, while concrete work can be priced on a per-cubic-yard basis.

Equipment costs can sometimes be estimated with surprising accuracy simply on the basis of their weight. For this purpose equipment can be

Table 33A Typical Cost-Capacity Factors for Different Types of Facilities

Type of Facility	Cost-Capacity Factor	Units of Capacity
Acetylene plant	0.73	tons/day
Aluminum plant	0.76	tons/year
Ammonia plant	0.72	tons/day
Steam boiler plant	0.75	pounds/hour
Cement plant	0.86	tons/day
Chlorine plant	0.62	tons/day
Electric generating plant (nuclear)	0.68	megawatts
Electric generating plant (steam)	0.79	megawatts
Industrial building	0.67	square feet
Municipal incinerator	0.80	tons/day
Oxygen plant	0.72	tons/day
Public housing project	0.75	number of rooms
Refrigeration system (mechanical)	0.70	tons
Sewage treatment plant (primary only)	0.68	gallons/day
Sewage treatment plant (primary and secondary)	0.75	gallons/day
Storage tank	0.63	gallons
Sulfuric acid plant	0.67	tons/day
Utility distribution main (gas and water)	0.91	pipe diameter
Utility distribution main (gas and water)	0.82	length installed

grouped into three broad categories: (1) precision, (2) mechanical/electrical, and (3) functional.

Precision equipment may be electronic or optical and generally costs about 10 times as much per pound as mechanical/electrical equipment which in turn costs about 10 times as much per pound as functional items. In the functional class items such as large power tools, automobiles, engine-generator sets, and heavy construction equipment are all found to cost approximately the same on a per-pound basis. Thus the costs of a 50,000-lb motor grader, a 70,000-lb crawler tractor, and a 90,000-lb pipelayer can all be estimated with reasonable accuracy (within 30 percent) by comparison with the cost per pound of an ordinary passenger automobile. Similarly, the cost of mechanical equipment —where a large part of the product's weight is made up of either electrical or mechanical parts—does not vary too much from the cost per pound of conventional small home kitchen appliances. To acquire a rough feel for precision equipment costs, the estimator need only check the cost per pound of items such as cameras and small electronic testing equipment.

In all types of order-of-magnitude estimates, the accuracy achieved is largely a function of the estimator's judgment and experience; he must be able to visualize the work as it will be done. Any projects identified as being potentially favorable can then be appraised in greater detail through more precise estimating methods.

SEMIDETAILED ESTIMATES

Semidetailed, conceptual, or budget estimates, should be accurate to within about 10 percent of the actual project cost. This level of accuracy is usually adequate for making decisions regarding project feasibility—whether or not the owner decides to proceed with the project. The engineer's estimate on new construction work might be considered a semidetailed estimate in most cases. The accuracy of the semidetailed estimate depends on the amount and quality of information available at the time the estimate is made.

More information is required for making semidetailed estimates than for making order-of-magnitude estimates. Instead of using mathematical relationships between historical costs and estimated costs on proposed work, the new project must be considered on its own. Actual quotations should be obtained on major equipment and material items; some design data are necessary for making rough takeoffs; and approximate unit costs may be applied to the measured units.

The semidetailed estimates can be used to advantage by the engineer in several ways. Since accuracy is generally sufficient to warrant the authorization of project by an owner, the semidetailed estimate may be all that is needed, so long as the owner is aware of the accuracy limitations. This type of estimate is also useful in providing a rough check on detailed estimates obtained through more refined methods.

DETAILED ESTIMATES

Detailed estimates are used as a basis for making bids. These estimates should be accurate within 5 percent, having been prepared from complete engineering specifications, drawings, and site surveys. They are, however, time-consuming and costly to prepare and should be used only when absolutely necessary. In many cases, the contractor is the only one capable of making a good detailed estimate.

Considerable information is required for a detailed estimate, often more than is available at the time the engineer's estimate must be made. In preparing the estimate the specifications and plans must be studied carefully, quantity takeoffs made, prices obtained, labor availability and wage rates checked, subcontractors' estimates requested, and schedules made up. There are, in every detailed estimate, many opportunities to make mistakes, usually resulting in estimates that are too low. The most common errors leading to low estimates include the omission of items, undermeasurement of quantities, and underestimation of labor requirements.

CAPITAL VERSUS OPERATING COSTS

Typically, an engineering economy study compares a high-investment, low-operating-cost project with a low-investment, high-operating-cost alternative. It is therefore important in comparing these alternatives that both capital and operating costs be estimated with the same degree of accuracy.

The three basic types of estimating methods can be applied equally well to both capital and operating costs, their applicability depending on the purpose of the study. Either the order-of-magnitude or the semidetailed budget estimate is probably adequate for most engineering economy studies.

CAPITAL COST ESTIMATES. The total capital requirements of a new manufacturing plant can be broken down into the following components for estimating purposes.

1. Depreciable investment.
   a. Buildings and utilities.
   b. Equipment, including installation.
   c. Other.

2. Expensed or amortized investment.
   a. Research and development.
   b. Engineering.
   c. Startup.
   d. Other.
3. Nondepreciable capital requirements.
   a. Land.
   b. Working capital.
      i. Cash.
      ii. Receivables.
      iii. Inventory.

Additional breakdowns can be made of many of these items if the results appear to justify further detail. But unless there is some compelling reason to introduce complexity into the study, a relatively simple and straightforward treatment is preferable.

OPERATING COST ESTIMATES. Operating costs, or in the case of an industrial plant, production or manufacturing costs, can be divided into two broad categories: unit costs and period, or time, costs.

1. Unit costs.
   a. Materials.
   b. Labor.
   c. Variable overheads.
2. Period (time) costs.
   a. Fixed overheads.
   b. Capital charges.

Unit costs are estimated on a cost per unit of production or output basis and can generally be treated as being linear over a wide range of production volume. Period costs are incurred at a fixed level over a prescribed range of output since they are related to the level of investment rather than to the level of production.

A different type of cost breakdown is necessary in the case of a commercial, as opposed to an industrial, venture; but the *types* of costs—unit and period, or fixed and variable—are basically the same.

HOW TO MAKE QUICK MANUFACTURING COST ESTIMATES

There are many instances in which a high level of precision is neither necessary nor possible in manufacturing cost estimates and in which the available data are of limited accuracy or doubtful reliability. Sometimes no data at all are available.

In situations such as these, any output may be considered better than no output, and the quickest, roughest sort of figures are gratefully accepted. The approach described here provides the quickest, roughest sort of figures.

SOURCES OF DATA. When no data are available that pertain directly to the project in question, the engineer must turn to the literature for his economic information. Fortunately, a considerable amount of financial and operating data are available for all types of manufacturing industries from readily accessible published sources. For example, the U.S. Department of Commerce publishes annually its *Survey of Manufactures* which gives general statistics for several hundred different industries and industry groups. The data reported in the survey cover labor, materials, and other costs, as well as the total value of shipments for each industry. These data are particularly useful as a guide to the overall operating cost structure of manufacturing industries.

Other sources, such as *News Front* and *Dun & Bradstreet*, tabulate and publish industry data from the financial records of major United States corporations. They summarize sales, assets, net worth, profit, and other useful financial data both by industry group and by individual companies.

If a new venture is being planned in some unfamiliar manufacturing line, a logical first step in the economic evaluation is to examine the financial and operating characteristics of individual companies involved in that field of manufacturing. Putting together the information that can be easily obtained from published sources can quickly give a reasonably clear picture of what a new venture might expect in any given field.

In making quick manufacturing cost estimates, all costs are related to annual sales. After the approximate scale of operations has been established, the expected annual sales can be estimated. For example, if a new plant were being considered to produce crushed stone for highway use at a rate of 500,000 tons per year, and if it were determined that the stone would bring a net price of $2.00 per ton, then the plant would generate net sales of $1,000,000 annually. From just this much information, a preliminary economic picture can be drawn of the operation.

CAPITAL REQUIREMENTS. Table 34 shows the total assets and net worth per dollar of annual sales for the 20 major manufacturing industry groups. Crushed stone falls in Standard Industrial Classification (SIC) category 32—stone, clay, and glass products—in which the total investment is in the neighborhood of $1.00/dollar of annual sales. If the proposed crushed stone plant were assumed to be typical of that industry group, the total investment would be in the same neighborhood, or about $1,000,000.

PRODUCTION COSTS. Table 35 summarizes the breakdown of production costs on a per-dollar-of-sales basis for the same 20 manufacturing industry

Table 34 Financial Characteristics of Major Industry Groups

SIC	Industry Group	Average Economic Life (years)	Total Assets per Dollar of Annual Sales ($)	Net Worth per Dollar of Annual Sales ($)
20	Food products	12	0.48	0.27
21	Tobacco	15	0.69	0.38
22	Textile mill products	14	1.04	0.54
23	Apparel	9	0.62	0.29
24	Lumber and wood products	10	0.99	0.54
25	Furniture and fixtures	10	0.63	0.41
26	Paper products	14	0.99	0.54
27	Printing and publishing	11	0.84	0.46
28	Chemicals	11	0.89	0.52
29	Petroleum products	16	1.18	0.72
30	Rubber and plastics	12	0.84	0.40
31	Leather	11	0.56	0.30
32	Stone, clay, and glass	15	1.00	0.61
33	Primary metals	16	1.15	0.64
34	Fabricated metal products	12	0.72	0.40
35	Nonelectrical machinery	12	0.90	0.52
36	Electrical equipment	10	0.69	0.35
37	Transportation equipment	12	0.63	0.33
38	Instruments	12	0.85	0.54
39	Miscellaneous manufacturing	12	0.70	0.43
	All manufacturing (average)	12	0.85	0.46

Table 35 Operating Characteristics of Major Industry Groups

SIC	Industry Group	Cost per Dollar of Annual Sales: Direct Labor ($)	Direct Materials ($)	Indirect Payroll ($)	All Other ($)
20	Food products	0.08	0.65	0.05	0.22
21	Tobacco	0.07	0.55	0.02	0.36
22	Textile mill products	0.20	0.56	0.05	0.19
23	Apparel	0.23	0.48	0.07	0.22
24	Lumber and wood products	0.21	0.52	0.05	0.22
25	Furniture and fixtures	0.24	0.45	0.09	0.22
26	Paper products	0.17	0.51	0.07	0.25
27	Printing and publishing	0.20	0.33	0.16	0.31
28	Chemicals	0.09	0.43	0.07	0.41
29	Petroleum products	0.04	0.75	0.02	0.19
30	Rubber and plastics	0.20	0.45	0.08	0.27
31	Leather	0.25	0.46	0.07	0.22
32	Stone, clay, and glass	0.21	0.41	0.08	0.30
33	Primary metals	0.17	0.56	0.06	0.21
34	Fabricated metal products	0.21	0.46	0.09	0.24
35	Nonelectrical machinery	0.21	0.42	0.12	0.25
36	Electrical equipment	0.19	0.42	0.14	0.25
37	Transportation equipment	0.15	0.57	0.08	0.20
38	Instruments	0.17	0.34	0.14	0.35
39	Miscellaneous manufacturing	0.21	0.44	0.10	0.25
	All manufacturing (average)	0.16	0.51	0.10	0.23

groups. Here it can be seen that direct labor and materials costs account for 62 percent of the total sales dollar, while indirect payroll costs come to 8 percent of sales. This leaves 30 percent to cover fixed costs such as depreciation, interest, insurance, property taxes, and general administrative expenses. Table 36 summarizes the operating economics of the proposed crushed stone plant, again using the SIC-32 industry averages.

As shown in Table 36, direct costs of $620,000 for labor and materials leave $380,000 to cover fixed costs. Deducting administrative or indirect payroll expenses of $80,000 leaves $300,000.

Fixed costs are related primarily to fixed investment, with depreciation and interest the most important elements. The average economic life for SIC-32 industries is 15 years (see Table 34); thus the average straight-line depreciation rate is 6.67 percent. The cost of capital for a venture such as this may be about 9.0 percent. By referring to a set of interest tables, the

Table 36 Example of Preliminary Cost Estimate for Crushed Stone Operation

Capacity: 500,000 tons/year	
Net product price: $2.00/ton	
Net annual sales: 500,000 tons at $2.00/ton	$1,000,000
Total investment	1,000,000
Direct production costs at 62% of annual sales	620,000
Gross operating profit	$ 380,000
Indirect costs	
Indirect payroll at 8% of annual sales	$ 80,000
Average depreciation at 6.67% of investment	67,000
Average interest at 5.74% of investment	57,000
Other indirect expenses at 2.5% of investment	25,000
Total indirect costs	$ 229,000
Net profit before income taxes	$ 151,000
Income taxes at 50% of pretax net profit	75,000
Net profit after taxes	$ 76,000
Plus depreciation	67,000
Net annual cash flow	$ 143,000

capital recovery factor for a 15-year, 9.0-percent investment is found to be 0.1241, or 12.41 percent per year. Since 6.67 percent has been established as the average depreciation rate, the balance of the 12.41-percent capital recovery rate represents the average interest: 12.41 − 6.67 = 5.74 percent annually for interest charges over the life of the project.

Insurance and property taxes are generally about 1.5 percent of the fixed investment, and miscellaneous administrative and general investment-related expenses can be included at 1.0 percent annually.

In the example fixed costs total $229,000, leaving a net profit before taxes of $151,000. Deducting income taxes at a 50-percent rate results in a net profit after taxes of $76,000. Adding the depreciation back in gives a net annual cash flow of $143,000, or 14.3 percent of the total investment. Again referring to interest tables, it is found that a capital recovery rate of 14.3 percent (or 0.143) represents a net return of about 11.5 percent annually over a 15-year period. Since the cost of capital was estimated at 9.0 percent, the proposed venture could be considered—at this early stage—economically feasible.

USEFULNESS OF THE ESTIMATES. Obviously, many refinements can be made in this rough estimate as better data become available. A far truer picture can in fact be drawn even from published literature. The important point is simply that a rough but useful estimate can be made quickly and easily, without any specific data, to test the reasonableness of a proposed manufacturing operation. Then, when better information becomes available, it can replace the general industry standard figures until a reasonably accurate estimate is ultimately obtained.

ESTIMATING THE ACCURACY OF ESTIMATED COSTS

A lot of money is either spent or committed on the basis of an engineer's estimate. Often even the engineer who made the estimate is uncertain about the accuracy of his estimate.

The engineer's estimate is usually made up of several uncertain and often unpredictable elements. These elements are then either multiplied or added together, or both, to come up with a final lump-sum figure. Under such conditions it is surprising that most engineer's estimates are as accurate as they are.

The next best thing to being certain about an estimate's accuracy is to be certain about just how uncertain the estimate is. Adding numbers together increases the accuracy of their total, while the other arithmetic operations—subtraction, multiplication, and division—mangify any inaccuracies present in the individual numbers. If enough separate items can be added together, fairly good accuracy can often be attained without being especially precise on individual items.

ACCURACY OF ENGINEER'S ESTIMATES. A comparison of engineer's cost estimates and actual contract award prices on nearly 100 recent projects was made, drawing from the projects reported over the past year in the "Unit Prices" section of *Engineering News-Record.* The jobs reported ranged in size from $119,000 to more than $15 million and involved anywhere from 2 to 13 bids. The results are summarized in Table 37.

Only 3 percent of the engineer's estimates were more than 20 percent below the low bidder's price; similarly, just 3 percent of the estimates were off by more than 20 percent on the high side. Some 30 percent of the engineer's estimates were within 5 percent of the actual contract award price; just over half were within 10 percent; about three-fourths were within 15 percent; and 94 percent of the estimates came within ±20 percent.

ACCURACY OF PRODUCTS. When two or more independent variables are multiplied together, any inaccuracies in the individual variables are amplified

Table 37 Accuracy of Engineer's Estimates

Ratio, Low Bid to Engineer's Estimate	Percent of Contracts Awarded
0.800 or less	3.0
0.801 - 0.850	15.2
0.851 - 0.900	19.7
0.901 - 0.950	10.6
0.951 - 1.000	15.2
1.001 - 1.050	15.2
1.051 - 1.100	10.6
1.101 - 1.150	4.5
1.151 - 1.120	3.0
1.201 or more	3.0
	100.0

in their product. The mathematical expression for a two-variable relationship, say range ($A \times B$), is:

$$\sqrt{A^2 \times \text{range}^2\,(B) + B^2 \times \text{range}^2\,(A)}$$

where the range of the variable is expressed in absolute units, not as a percent. For example,

$$\begin{aligned}
& (100 \pm 10\%) \times (60 \pm 20\%) \\
&= (100 \pm 10) \times (60 \pm 12) \\
&= (100 \times 60) \pm \sqrt{100^2 \times 12^2 + 60^2 \times 10^2} \\
&= 6000 \pm \sqrt{1{,}440{,}000 + 360{,}000} \\
&= 6000 \pm 1340 \\
&= 6000 \pm 22.4\%
\end{aligned}$$

Table 38 shows the percent range that can be expected in the product of two independent variables, depending on their respective accuracies (or inaccuracies).

Table 39 illustrates the use of the information in Table 38 in a typical engineering cost estimate. For example, in the rock excavation estimate, the quantity (accurate within ±15%) is multiplied by the unit cost (which is accurate within ±30%) to arrive at the estimated total cost for that item, which is found (from Table 38) to be accurate within ±33.6 percent.

Table 38 Percent Range Expected in Product of Two Independent Variables

Percent Range of Variable A	Percent Range of Variable B						
	0	5	10	15	20	25	30
0	0	5.0	10.0	15.0	20.0	25.0	30.0
5	5.0	7.1	11.2	15.8	20.6	25.5	30.4
10	10.0	11.2	14.1	18.0	22.4	27.0	31.6
15	15.0	15.8	18.0	21.2	25.0	29.2	33.6
20	20.0	20.6	22.4	25.0	28.3	32.1	36.1
25	25.0	25.5	27.0	29.2	32.1	35.4	39.1
30	30.0	30.4	31.6	33.6	36.1	39.1	42.5

Table 39 Accuracy of the Product of Several Inaccurate Figures

Item	Estimated Quantity	Estimated Unit Cost	Estimated Total Cost
Rock Excavation	6,000 CY ± 15%	$5.75 ± 30%	$34,500 ± 33.6%
Common Excavation	15,000 CY ± 10%	$2.75 ± 30%	$41,250 ± 31.6%
Embankment	30,000 CY ± 10%	$1.50 ±25%	$45,000 ± 27.0%
Cement	35,000 bbls ± 5%	$4.90 ± 10%	$171,500 ± 11.2%
Reinforcing Steel	50 tons ± 5%	$325 ± 5%	$16,250 ± 7.1%

Division works exactly the same way as multiplication, increasing the range of inaccuracy. In multiplication or division the answer is less accurate than the least accurate of the individual numbers.

ACCURACY OF SUMS. Fortunately, the accuracy lost in multiplying and dividing can often be regained in adding. While multiplication and division decrease accuracy, addition improves accuracy. Mathematically, range $(A + B) =$

$$\sqrt{\text{range}^2\,(A) + \text{range}^2\,(B)}$$

where the range is again expressed in absolute units. Another example is:

$$\begin{aligned} &(100 \pm 10) + (60 \pm 12) \\ &= 160 \pm \sqrt{10^2 + 12^2} \\ &= 160 \pm 15.6 \\ &= 160 \pm 9.8\% \end{aligned}$$

In this case the expected error range of the total is less than the error in either of the individual numbers.

If one number were subtracted from the other, the expected range (in units) would be the same as if they were added, but the percentage range would be far greater. Using the same example once again:

$$\begin{aligned} &(100 \pm 10) - (60 \pm 12) \\ &= 40 \pm 15.6 \\ &= 40 \pm 39\% \end{aligned}$$

Table 40 Accuracy of the Sum of Several Inaccurate Figures

Item	Estimated Total Cost	Accuracy of Estimate Percent	Dollars	(Dollar Range)^2
Rock excavation	$ 34,500	± 33.6%	± $11,600	124,600,000
Common excavation	$ 41,250	± 31.6%	± $13,040	170,000,000
Embankment	$ 45,000	± 27.0%	± $12,150	147,600,000
Cement	$171,500	± 11.2%	± $19,210	369,000,000
Reinforcing steel	$ 16,250	± 7.1%	± $ 1,150	1,320,000
	$308,500	± 9.2%	± $28,300	812,520,000

Table 40 shows how the accuracy of the sum of the five estimated items from Table 39 is improved over their individual accuracies. Adding the five items having individual accuracy ranges of ±33.6, 31.6 27.0, 11.2, and 7.1 percent results in a total that would be expected to be accurate within 9.2 percent. This "principle of compensating errors" largely explains the accuracy of engineers' estimates, even when accurate estimating data are not available at the time the estimates are made.

SUMMARY

Cost estimates provide one of the most important inputs for economic evaluations. While the precision coming out of an economic analysis can be no better than the accuracy of the cost data going into it, there are many situations in which a high level of precision is neither necessary nor possible; short-cut estimating methods may be used to advantage in such cases. Order-of-magnitude estimates, accurate to within ±30 percent, are especially useful in preliminary studies; they may be based on historical cost-capacity relationships or use equipment ratios, physical dimensions, or weight as the basis for estimating costs. Semidetailed estimates are used in making budget decisions and should be accurate to within about 10 percent. The detailed estimate, accurate within 5 percent, is usually necessary only for contractors making fixed-price bids. In any cost estimate both capital and operating costs should be established with comparable precision, with the degree of detail in the cost breakdown depending on the purpose of the estimate. For the very quickest and roughest cost estimates on new ventures in which specific information is unavailable, general industry economic data can be used initially and refined later as better data become available. When the individual numbers in an estimate are relatively inaccurate, arithmetic operations such as subtraction, multiplication, and division increase the inaccuracies. However, adding numbers together improves the accuracy of their total, and fairly good overall precision can be obtained in an estimate made up of a large number of relatively inaccurate individual items.

10

Equivalent Annual Costs

All businesses incur two main types of costs: fixed and variable. Fixed costs are associated with the firm's capacity to do business, are related to its capital investment, and accrue with time rather than with the level of operation. Variable costs increase or decrease more-or-less directly with the firm's output of goods or services.

INDUSTRY COST STRUCTURES

Some industries are characterized by a high capital investment in fixed assets. Other industries can operate on relatively little fixed capital, but to do this generally requires a high variable cost rate. Table 41 shows the capital requirements and fixed-variable cost relationships typical of several selected major industry groups.

The rail and motor freight transportation fields offer a good example of two competing fields that operate with almost opposite capital-operating cost structures.

Railroads have a high investment per dollar of sales ($2.68), while truck lines have a far lower investment requirement in fixed assets and a corre-

Table 41 Cost Structure of Selected Industry Groups

Industry	Total Assets As Percent of Annual Sales	Costs As Percent of Sales	
		Fixed	Variable
Agriculture	89	33	67
Mining	145	42	58
Construction	83	32	68
Manufacturing	85	33	67
Railroad transportation	268	60	40
Motor freight transportation	57	29	71
Air transportation	138	41	59
Telephone and telegraph	272	61	39
Public utilities	318	68	32
Wholesale trade	42	26	74
Retail trade	58	29	71
Food stores	20	23	77
Business services	89	33	67
Engineering and architectural services	43	26	74

spondingly lower fixed cost; about 60 percent of railroad revenues go to cover fixed costs, compared with the truckers' 29 percent. Direct operating costs for railroads, though, are far lower than for truck lines—just 40 percent of the sales dollar, as opposed to 71 percent. Freight shipped by rail, then, has high fixed and low operating costs which must be reflected in railroad rate schedules; truck shipments, however, incur low fixed costs and high operating costs which result in a different type of rate structure.

Other examples of high-fixed-cost, low-variable-cost industries include: electric and gas utilities, with more than \$3.00 invested for each dollar earned annually; telephone utilities, with a total investment of 2.7 times their annual revenues; and airlines and mining companies, whose average investment runs close to $1.40 per annual sales dollar.

Having a high level of fixed costs means that a firm must maintain a consistently high volume or utilization rate to operate successfully. A freight train or airplane contributes about the same costs over a given distance whether it travels full or half full, and a public utility's total costs vary little regardless of the cubic feet of gas or kilowatt-hours of electrical energy sold. In all these cases, though, sales receipts or revenues vary almost directly with volume.

In low-capital-cost, high-variable-cost industries, just the opposite is true. If most costs are of a variable nature, reduced sales can be met with cutbacks in production, which in turn will be reflected in lower costs.

At high levels of production, fixed-cost-oriented industries have a distinct economic advantage over their variable-cost-oriented competitors; during recessions high-variable-cost-structured firms gain the advantage through their ability to cut costs.

For these reasons high capital cost structures are usually associated with industries or firms whose products or services benefit from a stable or at least predictable demand.

SUBSTITUTION OF CAPITAL FOR LABOR

Even within a single industry, management has many options regarding the firm's capital-operating (or fixed-variable) cost structure. Over a wide range of operations, capital can be substituted for labor, thus substituting fixed costs for variable costs. In the extreme case in which capital assets replace *all* production labor, a firm operates with fixed costs only, exclusive of raw materials. Its costs of doing business in this way are therefore almost uniform from year to year. This in fact has almost been accomplished in the petroleum refining field in which direct labor costs now account for less than 5 percent of total sales; raw materials are the only important variable costs remaining.

Regardless of the industry or type of business, a $5 capital investment can reduce direct labor costs by about $1. This relationship partially explains why the substitution of capital for labor is not necessarily desirable from the company's standpoint in many industries. If higher variable costs can be compensated for by higher prices, as is often the case, then it may be to a firm's advantage to retain the operating flexibility associated with a high variable cost rate and a minimum of fixed charges. The same is true in industries in which technological change occurs rapidly, accompanied by asset obsolescence. Even in fields in which high overheads might not be objectionable in view of the lower direct production or operating costs, the annual fixed charges on capital—typically in the 15–20 percent range—mean that little or no real economy is attained by the capital-labor substitution.

In order to obtain a realistic evaluation of the relative merits of various capital-operating cost alternatives, it is necessary to express all costs—both capital and operating—on an equivalent annual basis. Then a simple comparison of equivalent annual costs quickly establishes the optimum cost structure for an industry, firm, or new venture.

THE AMORTIZATION CONCEPT

Most engineering economy studies require a choice between alternatives. Since most engineering proposals involve expenditures and receipts of money over a period of time, an equitable basis is needed for comparing alternatives whenever the time periods or amounts differ. A convenient way to do this is by taking all future amounts and converting them to a single figure—the equivalent annual cost.

The equivalent annual cost of a time series of unequal payments is simply the uniform amount that would have to be set aside annually to have exactly the same economic effect as the unequal series of payments. Or, stated another way, it represents an annuity whose present value is equal to the present value of the alternative plan over the same time period and at the same interest rate.

For example, an individual holding a $15,000, 20-year, 5-percent mortgage on his home will spend $99 monthly over the 20-year period to cover the principal and interest on this loan. This $99 monthly payment for 20 years is, economically speaking, equivalent to a $15,000 cash outlay now, or to a single $39,800 payment 20 years from now, or to a $1204 payment at the end of each of the 20 years. These four alternatives all have exactly the same present value—$15,000—and the $1024 figure is said to be the equivalent annual cost for all three of the other situations indicated.

The equivalence of these four spending plans is not immediately apparent from the absolute figures alone; for equivalence depends not only on the amounts involved but also on the timing of those amounts and on the interest or discount rate. A change in any one of these three factors changes the equivalence relationships.

PROBLEMS OF EQUIVALENCE

All types of firms must choose between assets having different initial costs and offering different periods of useful service. Problems of equivalence, for example, arise frequently in feasibility studies concerning public utilities in which the company's financial structure (and regulatory controls) imposes certain restrictions on how capital expenditures can be charged. Many of these problems can be quickly and easily resolved on an equivalent annual cost basis.

As pointed out previously, capital charges refer to costs that are directly attributable to the amount of capital invested and are largely independent of the amount of production or sales. The annual charges against a firm's capital assets may include six major items: depreciation, interest, *ad valorem* taxes, administrative and general expenses, ROI, and income taxes, although some of these factors may be omitted or included elsewhere in the company's accounts.

Depreciation, the largest of the six capital cost elements, represents the annual cost of recovering a capital expenditure through accounting deductions and is allocated by spreading the total initial cost over the asset's estimated useful service life. Depreciation is usually computed as a percentage of the initial cost in the sum of years digits and declining balance methods; in the straight-line method, it is a fixed percentage.

Interest, similar to depreciation, is related directly to the investment. If the investment is financed through outside borrowing, then interest charges should be included directly as a capital charge. If the investment is internally financed, then interest charges might show up instead as a cost of capital or ROI rate. The capital recovery rate or factor, often used in engineering economy studies, includes both depreciation and interest in a single figure.

Ad valorem taxes, or property taxes, are paid by utilities and other firms on the value of their property. While the tax rate and the tax base may vary from year to year, *ad valorem* taxes are generally computed as a percentage of the original cost of the property less accrued depreciation, also called the net cost or book value, or as a percentage of some "assessed" value.

Administrative and general (A & G) expenses, typically amounting to between 1.0 and 2.0 percent of the total investment, refer primarily to the expenses involved in acquiring the asset initially and in keeping track of it over its economic life. Administrative and general expenses are sometimes lumped in with general overheads or fixed costs rather than being charged against specific assets.

Return on investment is the annual cost of money, whether financed by equity or debt capital, or a combination of the two. Return for a utility firm is usually expressed as a percentage of the original cost, original cost less accrued depreciation, fair value, net worth, or some other rate-base standard. For industrial firms the return can probably be expressed as a percentage of total investment or net worth. Return represents the cost of the use of money, as opposed to depreciation expense which refers only to the recovery of the initial out-of-pocket acquisition cost.

Table 42 Annual Costs on Asset Having $10,000 First Cost and 5-Year Life

	Account Balance			Annual Charges on Capital				
Year	Jan. 1	Dec. 31	Average Balance	Depreciation	Property Taxes	ROI	Income Taxes	Total Annual Cost
1	10,000	8,000	9,000	2,000	180	540	450	3,170
2	8,000	6,000	7,000	2,000	140	420	350	2,910
3	6,000	4,000	5,000	2,000	100	300	250	2,650
4	4,000	2,000	3,000	2,000	60	180	150	2,390
5	2,000	0	1,000	2,000	20	60	50	2,130

Income taxes, both federal and state, are directly related to the firm's profits or to the ROI; thus they are indirectly related to the same base as the return.

A firm in a 50-percent tax bracket, for example, requires a gross pretax return of twice its projected after-tax net profit to allow for payment of income taxes.

The actual amounts of these capital charges may vary widely among different industries and in different regions of the country. Nevertheless, their effect is essentially the same in all instances.

COMPARISON OF ALTERNATIVES

In the following example an asset with an initial cost of $10,000 is to be depreciated on a straight-line basis over a 5-year period. The other capital charges property taxes, ROI, and income taxes—are assigned percentage rates of 2.0, 6.0, and 5.0, respectively, applied only against the average undepreciated balance each year. Table 42 summarizes the capital costs by years over the 5-year period.

The problem, then, is to convert this unequal 5-year flow of costs to an equivalent single figure which can be used as a basis for comparison with investment alternatives. This is done by taking the present value of the total annual cost for each year, using an appropriate discount rate—in this instance the company's net ROI (6.0 percent) as shown in Table 43.

Dividing the total present value of annual costs ($11,292) by the sum of the discount factors (4.213) gives the equivalent annual cost of the series of

Table 43 Present Value of Annual Costs, $10,000 Asset with 5-Year Life

Year	Total Annual Cost	6% Discount Factor	Present Value of Annual Cost
1	3,170	0.944	2,992
2	2,910	0.890	2,590
3	2,650	0.840	2,226
4	2,390	0.792	1,893
5	2,130	0.747	1,591
Total		4.213	11,292

payments—$2680. This figure can be easily checked as the calculations in Table 44 show.

The net economic effect, then—that is, the present value of all future costs—is the same in both cases. The equivalent annual cost figure can therefore be used as a basis for comparison of assets having similar functions but requiring perhaps that a larger initial investment be spread over a longer useful service life. An example of this situation is shown in Table 45; this asset, costing $15,000, is expected to last for 10 years. All costs are computed in the same way as in the previous example.

Converting this 10-year series of costs into the single equivalent annual figure requires discounting the same as before (see Table 46).

The resulting equivalent annual cost in this instance is $18,906 divided by 7.360, or $2570. This amount, discounted annually at the 6-percent rate over the 10-year period, has approximately the same net present value as the preceding nonuniform series; the calculations are shown in Table 47.

The asset costing $15,000 and having a 10-year life is therefore a slightly better choice than the 5-year asset costing $10,000 when all capital-related costs are considered. The higher-priced asset has an equivalent annual cost of $2570, compared to $2680 for the asset with a lower initial cost.

This approach can be employed in a wide variety of engineering economy studies and provides a valid, convenient, and easily understandable comparison between alternatives, especially when they differ only in their original cost and expected service life.

Table 44 Equivalent Annual Cost for $10,000, 5-Year Asset

Year	Equivalent Annual Cost	6% Discount Factor	Present Value of Equivalent Annual Cost
1	2,680	0.944	2,530
2	2,680	0.890	2,385
3	2,680	0.840	2,251
4	2,680	0.792	2,123
5	2,680	0.747	2,002
Total			11,291

Table 45 Annual Costs on Asset Having $15,000 First Cost and 10-Year Life

	Account Balance			Annual Charges on Capital				
Year	Jan. 1	Dec. 31	Average Balance	Depreciation	Property Taxes	ROI	Income Taxes	Total Annual Cost
1	15,000	13,500	14,250	1,500	285	855	713	3,353
2	13,500	12,000	12,750	1,500	255	765	638	3,158
3	12,000	10,500	11,250	1,500	225	675	563	2,963
4	10,500	9,000	9,750	1,500	195	585	488	2,768
5	9,000	7,500	8,250	1,500	165	495	413	2,573
6	7,500	6,000	6,750	1,500	135	405	338	2,378
7	6,000	4,500	5,250	1,500	105	315	263	2,183
8	4,500	3,000	3,750	1,500	75	225	188	1,988
9	3,000	1,500	2,250	1,500	45	135	113	1,793
10	1,500	0	750	1,500	15	45	38	1,598

Table 46 Present Value of Annual Costs, $15,000 Asset with 10-Year Life

Year	Total Annual Cost	6% Discount Factor	Present Value of Annual Cost
1	3,353	0.944	3,165
2	3,158	0.890	2,811
3	2,963	0.840	2,489
4	2,768	0.792	2,192
5	2,573	0.747	1,922
6	2,378	0.705	1,676
7	2,183	0.665	1,452
8	1,988	0.627	1,246
9	1,793	0.592	1,061
10	1,598	0.558	892
Total		7.360	18,906

LEASING

Leasing is an important way of financing major assets by means of periodic payments over the asset's useful life. By leasing, the beneficial use of property can be acquired for a specified period of time for an agreed payment, and therefore the benefits of ownership can be enjoyed with little or no capital outlay.

Leasing in effect substitutes an annual expense for a capital investment. Thus it offers some of the advantages of a high-capital, low-operating-cost venture without incurring the high initial investment. Leasing is a way to assure equivalent annual costs by a contractual arrangement.

If the property being leased is nondepreciable (such as land), the lease payment represents only interest on the property's economic value. When a depreciable property is leased, the rental payment must cover both principal and interest, so that the asset can be paid off by its owner, with interest, over the leasing period. When a lease covers both depreciable and nondepreciable

Table 47 Equivalent Annual Cost for $15,000, 10-Year Asset

Year	Equivalent Annual Cost	6% Discount Factor	Present Value of Equivalent Annual Cost
1	2,570	0.944	2,426
2	2,570	0.890	2,287
3	2,570	0.840	2,159
4	2,570	0.792	2,035
5	2,570	0.747	1,920
6	2,570	0.705	1,812
7	2,570	0.665	1,709
8	2,570	0.627	1,611
9	2,570	0.592	1,521
10	2,570	0.558	1,434
Total			18,914

property (land and a building, for example), the annual rate is a composite figure, reflecting both elements.

The lease-versus-buy alternative is frequently available to a company in the process of acquiring new fixed assets. Just a few of the many types of properties commonly leased include supermarkets, office buildings, airplanes, automobiles, railroad cars, office copiers, and computers. These properties have widely varying characteristics, yet all of them possess certain qualities that can offer benefits to both the lessor and the lessee in a suitable leasing arrangement.

Probably the most common reason for leasing rather than buying a property is to avoid the initial capital expenditure. However, debt financing can accomplish the same objective, spreading out the payments in equal amounts over the asset's useful life much the same as a leasing arrangement would.

While the entire rental payment on a lease is tax-deductible, debt financing can again produce the same effect; the interest portion of annual debt repayments is deductible, and the principal payments can be offset by depreciation charges.

Leasing a depreciable property, then, does not reduce a firm's capital expenditures; it only rearranges them and places them in a different account. Whether the capital is drawn from capital or operating accounts actually makes no difference in the true rate of return earned by the purchased or leased property, except as affected by the timing of depreciation charges and interest payments.

Land and other nondepreciable properties are a completely different situation. Here outright ownership offers no tax relief, and considerable capital may be tied up nonproductively. By leasing land the annual payments become tax deductible and the total funds committed are likely to be much lower than if debt financing were employed. Potential opportunities for capital gains through appreciation of land values are of course sacrificed if land is leased rather than owned outright. For this reason many companies prefer land ownership, even without immediate tax advantages.

In general, it is more economical for a firm to own depreciable assets and to lease nondepreciable properties. Table 48 shows the annual percentage lease rates on assets leased for various periods of time.

In the table the annual pretax return refers to the interest rate earned by the owner of the property being leased. Thus if the owner of a building wishes to earn a pretax return of 10 percent on his invested capital, he would charge 11.0 percent annually on a 25-year lease, or 13.1 percent annually on a 15-year lease. If a nondepreciable property were being leased, the annual rental needed to produce a pretax return of 10 percent would of course be 10 percent.

Typically, supermarkets and similar commercial buildings are leased to yield the property owner a percent or so more than his cost of debt financing.

Table 48 Annual Percentage Lease Rate for Asset Depreciated over Leasing Period

	Leasing Period (years)					
Pretax Return	5	10	15	20	25	30
6	23.7	13.6	10.3	8.7	7.8	7.3
7	24.4	14.2	11.0	9.4	8.6	8.1
8	25.0	14.9	11.7	10.2	9.4	8.9
10	26.4	16.3	13.1	11.7	11.0	10.6
12	27.7	17.7	14.7	13.4	12.8	12.4
15	29.8	19.9	17.1	16.0	15.5	15.2

After the leasing period is up, he will have a building on his hands, and his expectations for the future development of a particular neighborhood will undoubtedly be an important consideration.

The duration of a lease depends on the characteristics of the property. Equipment subject to a high rate of obsolescence leases for a short time at a high rate; electronic data processing equipment, for example, is usually priced to pay out in 2–4 years on short-term leases, since the market for outdated computers is not strong. Similarly, supermarkets seldom remain at a given location for more than 15 years, and abandoned buildings must often be demolished to restore the land value.

Making a sound decision regarding the terms of a leasing agreement, or even reaching a lease-versus-buy decision, is particularly troublesome when there are intangible considerations involved. While the mathematics of the transaction are straightforward enough, it is difficult to assign realistic values to these intangible factors.

SUMMARY

Fixed and variable costs, common to all businesses and industries in widely varying proportions, significantly affect a company's economic position and competitive strategy. Some businesses are characterized by high fixed costs and low operating costs, while others take the opposite position. In general, a $5 investment can reduce direct labor costs by about $1 annually, thus allowing management considerable flexibility in establishing a firm's overall cost structure. In identifying a desirable fixed-variable cost relationship for either a firm as a whole or a single new venture, or in comparing alternative projects having different cost characteristics, it is important to translate these two types of costs into a single, equivalent annual amount. For this purpose the amortization concept is useful. An initial capital expenditure can be converted to a series of annual payments by applying appropriate rates for depreciation, interest, *ad valorem* taxes, A & G expenses, ROI, and income taxes. Then the annual amounts can be converted to an equivalent annual basis by applying an appropriate discount rate to each annual payment, summing the present values of the future annual amounts, and dividing by the sum of the annual discount rates. As an alternative to outright ownership of capital assets, leasing offers a means of contractually obtaining equivalent annual costs by substituting a uniform yearly payment for an initial capital outlay.

11

Breakeven Analysis

Breakeven analysis is a useful tool in business and profit planning, since many management decisions require that the effect of changing sales volumes, markups, and overheads be viewed in terms of the company's overall profit position. One of the major problems encountered by management in making these types of decisions is being able to visualize correctly the interactions among all the different elements that must be considered.

Breakeven analysis can clearly illustrate the essential relationships among such factors as direct job costs, overhead costs, percentage markups, sales volumes, and profits. By clarifying these relationships, management can objectively evaluate the impact of changes in the different elements and can select appropriate strategies for working toward the firm's objectives.

BREAKEVEN TERMINOLOGY

In breakeven analysis, costs are broken down into their fixed, semivariable, and variable elements.

Fixed costs remain constant over a period of time regardless of work performed by the firm. Fixed costs include many of the general overhead items.

Semivariable costs may actually be fixed within certain relatively narrow ranges of sales volumes but generally increase in some relation to sales. Semivariable costs include some capacity-related general overhead items, plus any additional administrative expenses necessary to handle heavy work loads. If semivariable costs vary directly with the volume of work, they can be included with variable (or direct) costs; if they are stable over well-defined ranges, they may be handled in the same manner as fixed costs.

Variable costs, also referred to as *direct costs*, *job costs*, or *out-of-pocket costs*, are those that vary directly with the amount of work undertaken. Job-related overheads may be included in this category.

Total income and *total sales* are used synonymously, referring to the total amount of money generated by the business or project being considered.

Total costs are made up of the sum of all fixed, semivariable, and variable costs incurred over a specified time period.

Profit refers to the difference between total income and total cost whenever the income is greater than the cost. Unless otherwise specified, profit is computed before income taxes.

Loss refers to the difference between total cost and total income whenever the cost is greater than the income.

The breakeven point is the sales volume at which there is neither profit nor loss, or where total income is equal to total cost.

The breakeven point may also be expressed as a percent of capacity, or in terms of units of production, instead of as a dollar sales volume.

The *markup* is the percentage added to estimated direct costs (usually, direct labor and materials). The markup must be sufficient to recover all fixed costs and, hopefully, to return a profit on the invested time and capital.

INFORMATION REQUIREMENTS

Cost data provide the basis for breakeven analysis. These basic cost data should be compiled, analyzed, and related to different levels of sales volume.

Historical accounting records usually provide the most convenient sources of information used in breakeven analysis, although in the absence of such records good estimates may be satisfactory for planning purposes.

Table 49 shows a summary operating statement (or profit-and-loss statement) for a typical manufacturing company with annual sales of $1,000,000. By using only these basic data, a useful and meaningful breakeven analysis can be made.

Table 49 Operating Statement for Typical Manufacturing Company

Total sales		$1,000,000
Cost of operations		
Direct job costs	$670,000	
Fixed costs	230,000	
Total costs		900,000
Net profit before income taxes		$ 100,000

BASIC BREAKEVEN ARITHMETIC

From the fixed costs of $230,000 shown in the operating statement and the direct job costs figure of $670,000, representing all variable costs, this firm's break-even point can be quickly determined.

The breakeven point, by definition, is the point at which total income equals total cost. Total cost in turn includes both fixed and variable costs. To find the breakeven point for this firm, then, the amount of variable costs, which when added to the $230,000 fixed costs equals total income at that point, must be calculated. In making this calculation the total variable cost of $670,000 is assumed to have been accumulated at a constant rate over the $1 million sales volume; therefore for each dollar of sales, $670,000/$1,000,000, or $0.67, in variable costs are incurred. Or, stated another way, variable costs are equal to 67 percent of the sales volume.

Having accumulated the necessary information, the basic formula $V = F + D$ can be used to calculate the breakeven point, where V equals the volume of sales required to break even, F is the total amount of fixed cost, and D represents the total amount of direct costs at the breakeven point, expressed as a percentage of sales volume. In this case $0.67V$ can be substituted for D. Substituting the data from the preceding operating statement into the break-even formula then gives:

$$\$230{,}000 + 0.67V = V$$

$$0.33V = \$230{,}000$$

$$V = \$697{,}000$$

The breakeven point for this firm, then, occurs at a sales volume of $697,000. Should the company fail to achieve this volume a loss would result; and the

Table 50 Breakeven Volume for Typical Manufacturing Company

Sales volume at breakeven point	$697,000
Less variable expenses at 67% of sales	467,000
Amount remaining to cover fixed costs	$230,000
Less fixed costs	230,000
Net profit or loss	0

firm begins to realize a profit only after a sales volume of $697,000 has been reached. The breakeven calculation can be easily checked by working backward from the breakeven point, as shown in Table 50.

INDUSTRY BREAKEVEN STRUCTURES

In the preceding chapter typical cost structures were presented for several major industry groups. By using the fixed-variable cost relationships shown for these industries, some interesting comparisons can be drawn regarding their relative breakeven points.

When the variable cost rate for a firm is known (0.67 for the manufacturing firm in the preceding example) and the fixed costs can be identified for a specified sales volume, the basic breakeven formula $V = F + D$ can be restated as

$$V = \frac{F}{1 - R}$$

Again V and F represent the breakeven volume and fixed costs, and R is the variable cost rate.

This simple formula can be applied to the industry cost data to identify the breakeven points for each industry. This has been done in Table 51. An assumption was made that in all 14 of the industries a typical firm could realize a 10-percent pretax profit margin on a $1 million sales volume.

As shown in Table 51, the breakeven points for these industries range from a low of $565,000 for food stores with their low fixed costs and high-variable-cost rate, to a high of $853,000 for public utilities with just the opposite type of cost structure.

Some idea of the relative markups applied to direct costs in these industries can also be obtained from the data in Table 51. It is apparent that the high-

Table 51 Breakeven Point for Typical Firms in Selected Major Industry Groups*

Industry	Fixed Costs ($1000)	Variable Cost Rate	Breakeven Volume ($1000)
Agriculture	230	0.67	697
Mining	320	0.58	762
Construction	220	0.68	687
Manufacturing	230	0.67	697
Railroad transportation	500	0.40	833
Motor freight transportation	190	0.71	655
Air transportation	310	0.59	756
Telephone and telegraph	510	0.39	836
Public utilities	580	0.32	853
Wholesale trade	160	0.74	615
Retail trade	190	0.71	655
Food stores	130	0.77	565
Business services	230	0.67	697
Engineering and architectural services	160	0.74	615

*Assumes that the firm has a 10% pretax profit margin at $1,000,000 sales volume.

fixed-cost industries must apply higher percentage markups to their direct or out-of-pocket costs in order to recover their fixed costs. The average percentage markup (M) on direct costs is simply

$$M = \frac{100}{V} - 100$$

where V is the variable cost rate expressed as a decimal. For the 14 industries shown, markups range from a high of 213 percent for public utilities [(100/0.32) − 100], where the direct costs of energy production are far outweighed by the fixed costs of production and distribution, to a low of 30 percent for food stores [(100/0.77) − 100], where the purchased cost of goods sold represents a high proportion of the retail selling price.

THE BREAKEVEN CHART

While the arithmetic calculations involved in breakeven analysis are simple and should be thoroughly understood, the use of a breakeven chart provides a more vivid picture of exactly where the firm stands with respect to its breakeven point. Also, the relationships among the various factors that influence the breakeven point can best be expressed graphically. The breakeven chart, then, offers a convenient means of clearly illustrating several important cost relationships that would otherwise be quite difficult to express. Many feasibility reports could probably benefit substantially by substituting a few well-chosen charts for several thousand words.

The same information is required to construct a breakeven chart as is used in computing the breakeven point arithmetically. The conventional breakeven chart assumes that fixed costs remain constant regardless of the sales volume, and that variable or direct job costs change in direct proportion to sales.

Figure 25 shows the essential features of a breakeven chart for a manufacturing firm having total annual sales of $1,000,000; direct job costs make up $670,000 of this total, overhead accounts for $230,000, and $100,000 is left as net profit before taxes. This operation incurs variable costs at a rate of 67 percent of sales ($670,000/$1,000,000).

In Figure 25, the vertical axis represents both income and cost, and the horizontal axis shows total sales volume. The units of measurement on both scales are dollars.

At zero sales income is also zero. Therefore the total income line is a straight line passing through the origin, representing the income increasing as sales increase. Since the company's income is presumably derived solely from sales, the total income shown on the vertical axis is always equal to the total sales volume shown on the horizontal axis.

Next the total amount of fixed costs is plotted on the graph. Fixed costs in this case total $230,000 throughout the range of sales shown on the breakeven chart.

Variable costs are then added to fixed costs to arrive at total costs, with the slope of the variable cost line depending on the rate at which these costs are incurred; in this case at 67.0 percent of sales. These variable costs are

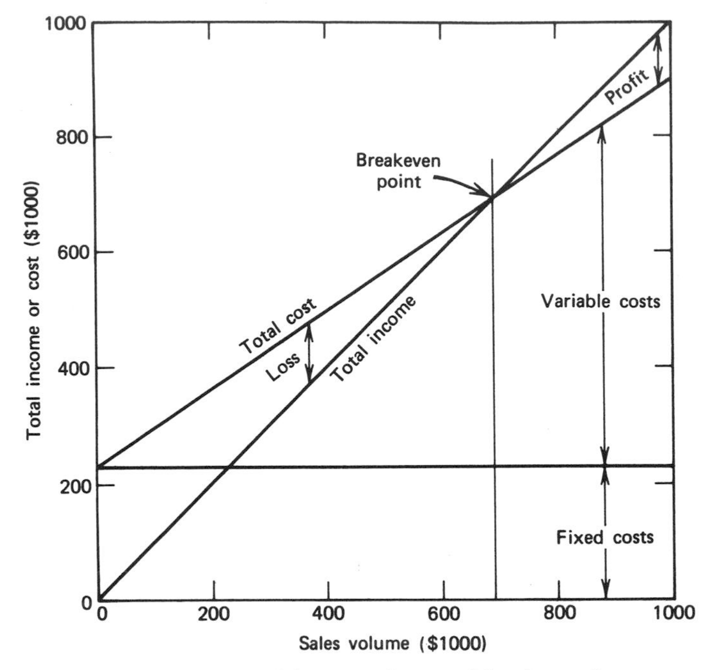

Fig. 25. *Essential features of a typical breakeven chart.*

plotted above the fixed cost line, thereby giving the total cost of operations for any given level of sales. The total cost line—the sum of fixed and variable costs—indicates the total cumulative amount spent at a specific sales volume.

The point at which the total cost line intersects the total income line—$697,000 in this example—is the breakeven point. To the left of this point, the vertical distance between the total income and the total cost lines indicates a net loss; to the right, it represents the net profit.

At the $697,000 breakeven point, sales revenues are exactly matched by the two elements of cost: $230,000 in fixed charges and $467,000 in variable costs. At a sales volume of $800,000, profits of $34,000 result; with sales of only $500,000, a net loss of $65,000 accrues.

The breakeven point can be described equally well in terms of the firm's operating capacity instead of as a sales volume, should this measure be more meaningful to management. For example, if the firm's annual capacity were $1,500,000, the breakeven point would occur at about 46.5 percent of capacity. To earn net profits of $34,000 would require operating at 53.3 percent of capacity. In either case the procedure is the same, differing only in the units of measurement along the horizontal axis of the graph.

EFFECT OF VARYING FIXED COSTS

In the preceding example fixed costs were assumed to remain constant regardless of the level of sales. In most cases, however, fixed costs are apt to increase as the firm's capacity to do business increases. Usually, these additional overhead costs are not incurred gradually but come in steps, with each step covering a sizable range—just as adding another generator set to a power plant increases plant cost in a single step, which then remains at the same level over the next range of generating capacity. Figure 26 illustrates the effect of varying fixed costs on a firm's breakeven point. The scales are the same as in the previous example. As indicated in Figure 26, the effect of lowering the level of fixed costs is to cause the total cost line to intersect the total income line more quickly, indicating a lower breakeven point. Lowering the fixed cost from \$230,000 to \$150,000 reduces the breakeven point from \$697,000 to \$455,000; thus an \$80,000 reduction in fixed costs results in a \$242,000 reduction in the breakeven point.

By reducing fixed costs by any given percentage, the breakeven point is reduced by that same percentage. Or by increasing the fixed costs by any given percentage, the breakeven point is increased by a like percentage.

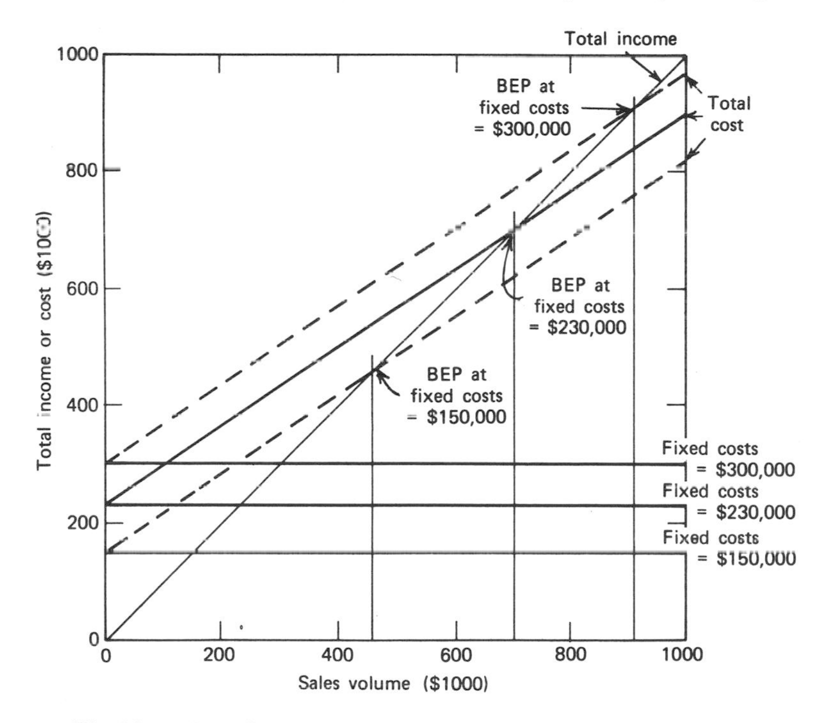

Fig. 26. *Effect of varying fixed costs on the breakeven point (BEP).*

So, increasing fixed charges from $230,000 to $300,000 moves the breakeven point farther away, from $697,000 to $909,000. This means that $212,000 in additional sales has to be generated just to make up for the additional $70,000 in fixed cost obligations. Again the same percentage increase must be made in sales as in fixed costs to retain the same position of economic equilibrium.

Figure 27 shows how the same company's operations might look if these three levels of fixed costs—$150,000, $230,000, and $300,000—are incurred over specific ranges of sales volume. In this figure fixed costs of $150,000 are associated with sales volumes of zero to $500,000; $230,000 is incurred at sales volumes between $500,000 and $1,000,000; and $300,000 in fixed charges is necessary to support sales between $1,000,000 and $1,500,000.

The first breakeven point for this operation occurs at sales of $455,000. Profits rise to $15,000 at a $500,000 sales volume, at which time additional fixed charges must be incurred to support any additional volume, and the breakeven point rises to $697,000. Breaking even at the higher overhead requires $242,000 more sales, and to make the same amount of profit as was made at the $500,000 volume ($15,000) now requires sales of $742,000—also $242,000 more than before.

The breakeven chart provides a useful visual representation or model of a firm's overall operating picture, showing important relationships between income and costs. It vividly emphasizes the sometimes alarming effect of overhead or fixed costs on the breakeven point. The effect of overhead costs

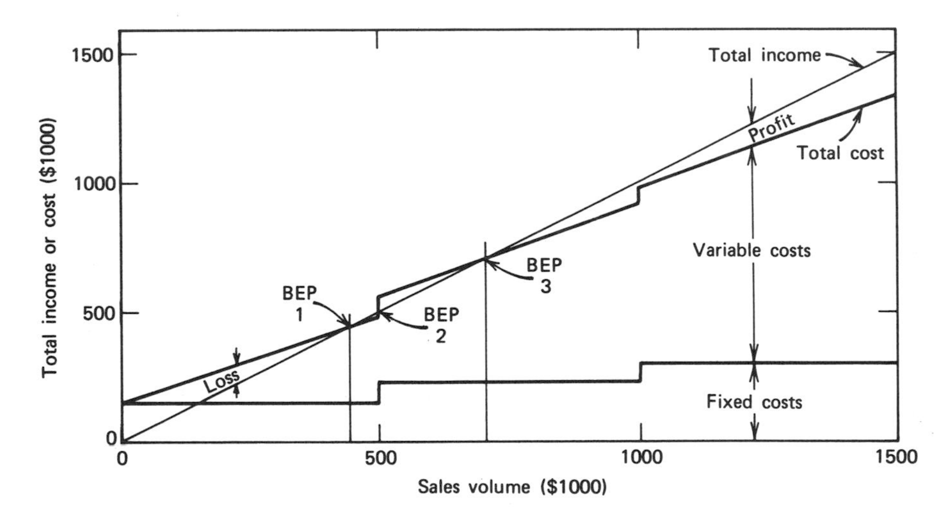

Fig. 27. *Effect of varying fixed cost levels over different volume ranges on the breakeven point (BEP).*

on the breakeven point depends primarily upon the rate at which variable costs are incurred; in some cases each change of $1 in fixed costs may change the breakeven point by $4 or more. Failure to consider the effect of fixed costs in a firm's pricing policies can, and often does, result in making a profit on every job, but still losing money in total.

EFFECT OF MARKUPS

Whether selling new construction, engineering services, or retail items, the seller's markup—the percentage he adds to his direct costs to cover overhead and profit—has an important effect on the volume of sales he must attain to break even. The higher the markup, the lower the breakeven point; the lower the markup, the higher the breakeven point.

Looking at the seller's direct cost, books, as a familiar example, usually carry about 40-percent discount to the retailer; in other words, the selling price of a $10 book is made up of the bookseller's cost ($10 less 40%, or $6) and his $4 markup. From the bookseller's standpoint, then this $4 markup represents a markup of two-thirds, or 66.7 percent, on his out-of-pocket cost. On this basis his direct costs—that is, the actual out-of-pocket costs of the books he sells—are accruing at a rate of 60 percent of sales. His problem is to sell enough books to cover his overhead and return a profit.

This problem is of course also typical of engineers, contractors, and businessmen in general.

A consulting engineering or other type of service firm (legal, accounting, advertising, etc.) might typically be pricing its services by applying a multiplier of from 2.0 to 3.0; these multipliers actually represent markups on direct costs of 100 and 200 percent. It is hoped that at these markups enough services will be sold to cover overhead and return a profit.

At the 100-percent markup (or 2.0 multiplier), the firm's direct costs are accruing at a rate of 50 percent of sales. At the 200 percent markup, the rate at which his variable costs are incurred is just 33.3 percent of sales.

In another line of business, the general contractor may be operating with a 10-percent markup on his direct, or variable, costs. Then variable costs are found to accrue at a rate of 90.9 percent of sales.

A few more fortunate subcontractors may be able to get by with a 25-percent markup on their direct costs; their variable cost rate, then, is 80 percent.

There is obviously a direct relationship between the percentage markup and the variable cost rate; a low variable cost rate is associated with a high markup and a low breakeven point, and a high variable cost rate with a low markup and a high breakeven point.

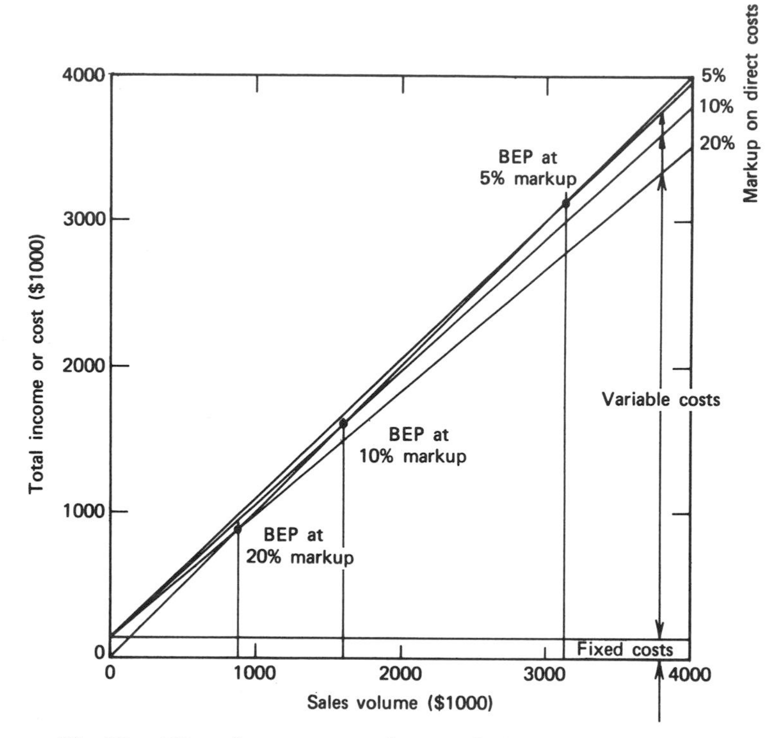

Fig. 28. *Effect of percentage markups on the breakeven point (BEP).*

Figure 28 illustrates how and why the breakeven point varies with the markup. In the example of Figure 28, the firm has annual fixed costs of $150,000. With a 10-percent markup, the breakeven point occurs at a sales volume of $1,650,000. This volume is matched by total costs of the same amount, made up of $1,500,000 in direct (or variable) costs plus the 10-percent markup that covers the $150,000 in fixed costs. The variable cost line has a slope of $909,000 in $1,000,000, or 90.9 percent of sales.

A higher markup decreases the slope of the variable cost line, causing it to intersect the total cost line more quickly. A 20-percent markup results in a breakeven point of $900,000, representing a slope of $833,000 in $1,000,000, or 83.3 percent of sales.

Similarly, a lower markup, by raising the slope of the variable cost line, causes it to intersect later with the total cost line, denoting a higher breakeven point. For a 5-percent markup, the breakeven point is at $3,150,000, made up of $2,857,000 in direct costs, a $143,000 markup, and $150,000 in fixed costs. Variable costs (including the markup) in this case are incurred at a rate of 95.2 percent of sales.

Such is the fate of the contractor. The bookseller, however, with his 66.7 percent markup, is able to recover his $50,000 in overhead on a sales volume of only $125,000; and the consultant, with a 200-percent markup, can break even with just $75,000 worth of business.

THE BREAKEVEN FORMULA

The mathematical relationship among fixed charges, percentage markups on direct job costs, and the volume of work required to break even can be easily derived. Algebraically, these relationships can be expressed as:

$$\frac{V}{F} = \frac{100 + M}{M} = \frac{100}{M} + 1$$

where V is the breakeven volume, F the fixed charges or overhead, and M the percentage markup on direct costs, expressed as a percent.

By using this formula, a 10-percent markup is found to require a sales-to-overhead ratio of 110/10, or 11.0. Similarly, a 20-percent markup requires a sales volume of 120/20, 6.0 times the overhead. A 100-percent markup results in a breakeven volume of 200/100, or 2.0 times the overhead. The relationship between the percentage markup on direct costs and the ratio of sales to fixed costs is shown graphically in Figure 29.

SIGNIFICANCE OF RELATIONSHIPS

Since a firm's breakeven point is a function of its overhead costs and percentage markups, any change in either of these factors is greatly amplified in the amount of work the firm must obtain to break even. Working on a 10-percent markup, each $1 of overhead must be matched by $11 in sales. This means that, if an additional $1000 of overhead expense is taken on, an additional sales volume of $11,000 must be generated just to maintain the same profit position. Reducing overheads by a like amount reduces the breakeven point by the same $11,000.

Important changes in a firm's financial performance can be brought about by varying markups and sales volumes. In highly competitive industries such as construction, it may not be possible to raise markups, but equally beneficial results may be obtained by reducing overheads. Firms engaged in selling unique services or products may, however, find it easier to increase profits by raising prices.

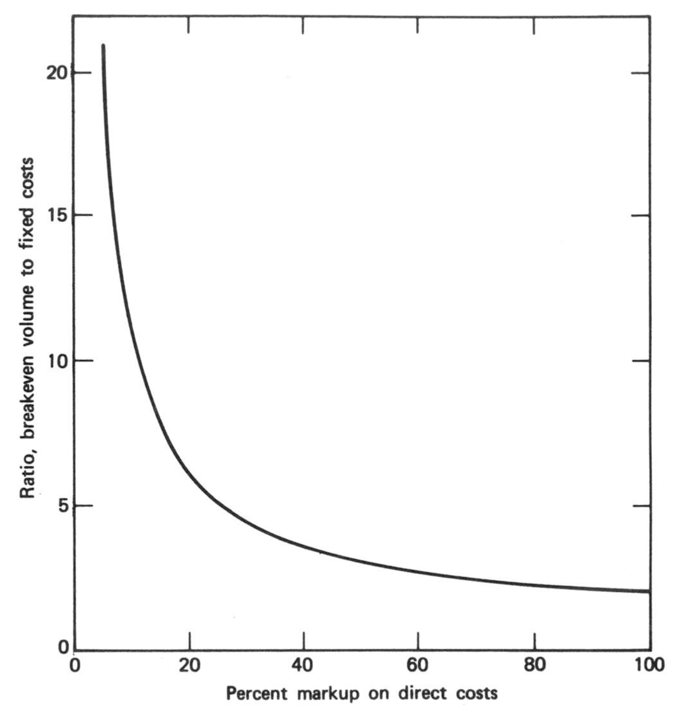

Fig. 29. *Relationship between percentage markups and the breakeven volume/fixed cost ratio.*

SUMMARY

Breakeven analysis is a useful tool in defining and describing the relationships between a firm's sales revenues, fixed and variable costs, markups, and profits. A company's breakeven point can be calculated from financial data usually available in its financial statements and varies widely according to the characteristics of the industry in which it operates. High-fixed-cost operations have relatively high breakeven points, while industries and companies operating with a high variable-cost rate have relatively low breakeven points. A company's breakeven point can be calculated from the formula $V = F/(1 - R)$, where V is the breakeven volume, F the fixed costs, and R the variable-cost rate expressed as a decimal. In addition to the variable-cost rate, fixed costs and percentage markups on variable costs are the primary determinants of a firm's breakeven point. A given percentage change in fixed costs changes the breakeven point by the same percentage and in the same direction. The mathematical relationship between fixed costs F, percentage markups on variable costs M, and the breakeven volume V can be expressed as $V/F = (100 + M)/M$. Thus a 10-percent markup requires sales of 11 times the fixed costs to break even, while a 20-percent markup lowers the sales/fixed costs ratio to 6.0. Overhead cost control therefore is particularly important in highly competitive industries such as construction and manufacturing, since each additional dollar of fixed cost incurred may be multiplied many times in the additional amount of sales income that must be generated to break even.

12

Financial Analysis for Profit Making

Making a profit is what business is all about. An adequate level of profits must return to business owners the cost of their personally contributed resources as well as a reward for their entrepreneurship and compensation for the risks involved.

The *adequacy* of profits is generally measured in terms of *profitability*, as a return on invested capital. However, profit analysis, as used here, deals primarily with the *magnitude* of profits and their relationship to other income account items.

FINANCIAL RATIOS FOR BUSINESS ANALYSIS

One of the best ways to judge the effectiveness of a company's management is to analyze its financial statements. The firm's management, too, can benefit greatly from the analysis, in spotting weaknesses and potential trouble spots before they become so serious that the financial condition is damaged beyond repair.

The two major financial reporting documents are the balance sheet and the operating (or profit-and-loss) statement. The balance sheet shows what the company's financial resources are as of a specific date, while the profit-and-loss statement describes—in dollars—the company's operations over a period of time. It is not the absolute size of the figures on these documents that are particularly meaningful to the financial analyst but rather the relationships among the different figures.

It must be remembered, however, that a corporation's balance sheet and profit-and-loss statements are "showpieces" prepared for widespread public distribution. The story told by these documents may appear to be quite different from what actually has happened. Consequently, footnotes and special situations cited in the reports should be carefully examined and their effects noted. Even with these shortcomings, analysis of the reported figures can be very useful and productive.

The importance of the manner in which income taxes are accounted for cannot be overemphasized in financial analysis. A company's pretax profit is an unrealistic figure in business, and it is therefore essential that the impact of taxes—both on the project in question and on the company's operation as a whole—be carefully evaluated before an important investment decision is made.

It should also be noted that industry profits are often of a cyclical nature. Not all industries are profitable at all times, and even during periods of strong economic activity the frequent occurrence of bankruptcies attests to the danger of assuming that profits are uniformly distributed. Good economic and financial data can be obtained on an industry-wide basis, and such data can be very useful as a basis for comparing the characteristics of individual companies. But only the data applying to a specific operation at a specific time can be used effectively in financial decision making.

Balance sheet and operating statement data for five selected corporations are summarized and presented in Tables 52 and 53. These data were taken from recent annual reports. Each of the five companies is among the leaders in its field, and would be considered a "blue-chip" investment.

In looking at these financial data, it is important to recognize that there is no such thing as an average company; and when dealing with a prospective customer or client, he should never be treated as an average customer—because he isn't. The wide variations in the balance sheet and operating statement accounts for these five leading firms should offer ample evidence in this respect.

Even when looking at a single company, it is not the absolute size of the figures on its financial statements that is particularly meaningful to the finan-

Table 52 Consolidated Balance Sheet Data for Selected Companies (Per Million Dollars of Annual Sales)

	Electrical Equipment Manfuacturer	Retail Department Stores and Mail Order	Chemical and Drug Producer	Integrated Petroleum Company	Retail Grocery Chain
Assets					
Current assets					
Cash	34,000	18,000	38,000	31,000	12,000
Receivables	172,000	26,000	160,000	204,000	4,000
Inventories	175,000	184,000	174,000	90,000	76,000
Other	10,000	10,000	55,000	50,000	7,000
Total current assets	391,000	238,000	427,000	375,000	99,000
Property, plant, and equipment	198,000	87,000	445,000	795,000	92,000
Other assets	88,000	32,000	66,000	144,000	3,000
Total assets	677,000	357,000	938,000	1,314,000	194,000
Liabilities					
Current liabilities					
Accounts and notes payable	90,000	104,000	110,000	150,000	39,000
Other	159,000	33,000	42,000	49,000	18,000
Total current liabilities	249,000	137,000	152,000	199,000	57,000
Long-term debt	88,000	38,000	101,000	137,000	10,000
Other liabilities	47,000	3,000	27,000	19,000	7,000
Total liabilities	384,000	178,000	280,000	355,000	74,000
Stockholders' equity					
Common and preferred stock	54,000	14,000	206,000	115,000	13,000
Capital surplus and retained earnings	239,000	165,000	452,000	844,000	107,000
Total stockholders' equity	293,000	179,000	658,000	959,000	120,000
Total liabilities and stockholders' equity	677,000	357,000	938,000	1,314,000	194,000

Table 53 Consolidated Operating Data for Selected Companies (Per Million Dollars of Annual Sales)

	Electrical Equipment Manufacturer	Retail Department Stores and Mail Order	Chemical and Drug Producer	Integrated Petroleum Company	Retail Grocery Chain
Net sales	1,000,000	1,000,000	1,000,000	1,000,000	1,000,000
Costs and expenses					
Cost of goods sold	876,000	709,000	512,000	585,000	785,000
Other costs and expenses	44,000	222,000	338,000	285,000	185,000
Total costs and expenses	920,000	931,000	850,000	870,000	970,000
Net profit before income taxes	80,000	69,000	150,000	130,000	30,000
Income taxes	37,000	35,000	68,000	27,000	15,000
Net profit after income taxes	43,000	34,000	82,000	103,000	15,000

cial analyst. The relationships among the different figures, though, can sometimes be quite revealing.

A company's current financial standing and past financial performance can be interpreted most conveniently when the relationships among various balance sheet and profit-and-loss statement items are expressed as ratios. The calculated ratios for a company can be used to compare the company's performance with the average performance of other businesses of the same type. Similar comparisons can also be made between a company's own financial statements over an extended time period, thus showing the firm's progress or lack of progress.

TYPES OF FINANCIAL RATIOS. There are three basic types of financial ratios. The first group includes balance sheet ratios, showing the relationship among various balance sheet items. The second type of ratio, called the operating ratio, relates expense accounts to income accounts on the profit-and-loss statement. The third group is composed of ratios involving both items on the balance sheet and items on the profit-and-loss statement. These three types of financial ratios can provide management with the basic information needed for effective control over the company's costs and finances.

Many different ratios can be derived from the balance sheet and profit-and-loss statement. In all, more than 30 different ratios are regularly compiled by various trade associations. Dun & Bradstreet, for example, publishes 14 key business ratios for more than 100 different businesses in retailing, wholesaling, manufacturing, and construction.

SIGNIFICANT RATIOS

Some ratios are important only in certain businesses, while others are of more general application. The following ratios are of significance in almost every type of business enterprise.

1. Net profit to net sales.
2. Net profit to net worth.
3. Net profit to working capital.
4. Net profit to total assets.
5. Net sales to net worth.
6. Salcs to working capital.
7. Sales to fixed assets.
8. Current assets to current liabilities.
9. Cash to current liabilities.
10. Fixed assets to net worth.

11. Current liabilities to net worth.
12. Total liabilities to net worth.

Table 54 shows the financial ratios for the five selected companies. These ratios were calculated from the balance sheet data in Table 52 and the operating data in Table 53.

The financial ratios shown in Table 54 differ substantially from those published by Dun & Bradstreet for the respective industries. Even the industry averages vary widely from year to year. Ratios, similar to the financial data from which they are derived, apply only to a specific company at a specified time.

NET PROFIT TO NET SALES. The net profit on sales, or profit margin, is an important financial yardstick. It indicates to some extent a company's competitive strength or its vulnerability to a decrease in either its sales volume or its profits. Usually, an increase in sales widens the profit margin, since fixed costs need not rise in direct proportion to sales. For the same reason profits tend to increase and decrease more rapidly percentage-wise than do sales. The profit margin also reflects the competitive situation within an industry. Grocery stores, for example, are noted for working on low profit margins. Among the individual companies surveyed, the grocery chain—as expected—showed the lowest margin, while the petroleum company was highest.

NET PROFIT TO NET WORTH. The net profit/net worth ratio is perhaps the most important of all financial ratios since it reflects the efficiency with which invested capital has been employed. A 10-percent return on invested capital is usually considered a bare minimum, and many well-managed companies earn more than 20 percent on their equity capital. The top 10 companies listed in *Fortune's* "500" generally earn better than 25 percent on their invested capital. All five of the sample companies showed better than a 10-percent return on net worth, with the department store operation realizing 19 percent. The return on net worth figure for a company gives a rough indication of what constitutes a *minimum* acceptable return on a new investment for that company. Certainly, any investment proposal indicating a return lower than what the company is presently earning can not be considered a particularly attractive investment opportunity.

NET PROFIT TO WORKING CAPITAL. Working capital represents the equity of owners in the company's current assets: the difference between total current assets and total current liabilities. This margin represents the "cushion" available to the business for carrying receivables and for financing day-to-day operations. The ratio of net profits to working capital is useful in measuring the profitability of firms whose operating funds are provided largely through borrowing, or whose permanent capital is unusually small in relation to volume

Table 54 Financial Ratios for Selected Companies

	Electrical Equipment Manufacturer	Retail Department Stores and Mail Order	Chemical and Drug Producer	Integrated Petroleum Company	Retail Grocery Chain
Net profit to net sales (%)	4.3	3.4	8.2	10.3	1.5
Net profit to net worth (%)	14.7	19.0	12.5	10.7	12.5
Net profit to working capital (%)	30.0	33.7	29.8	58.5	35.7
Net profit to total assets (%)	6.3	9.5	8.7	7.8	7.7
Net sales to net worth (ratio)	3.4	5.6	1.5	1.0	8.3
Net sales to working capital (ratio)	7.0	9.9	3.6	5.7	23.8
Net sales to fixed assets (ratio)	3.5	8.4	2.0	1.1	10.5
Current assets to current liabilities (ratio)	1.6	1.7	2.8	1.9	1.7
Cash to current liabilities (%)	13.6	13.1	25.0	15.6	21.0
Fixed assets to net worth (%)	97.6	66.5	77.8	97.9	79.1
Current liabilities to net worth (%)	85.0	76.6	23.1	20.7	47.5
Total liabilities to net worth (%)	131.0	99.4	42.6	37.0	61.6

of sales. Four of the five selected companies had profit/working capital ratios within a narrow range, with only the petroleum company falling outside the 29–36 percent area.

NET PROFIT TO TOTAL ASSETS. The ratio of net profits to total assets is closely related to the net profit/net worth ratio, except that here the theory is that return on investment should be measured in terms of *all* capital employed in the business—whether supplied in the form of equity or debt—and not in terms of equity interest only. All five companies surveyed earned between 6.3 and 9.5 percent on their total assets, even though their debt capital (as indicated in their balance sheets) may vary by as much as a factor of 10.

SALES TO NET WORTH. This ratio measures the rate of capital turnover, showing how actively the firm's capital is being put to work. If capital is turned over too rapidly, liabilities are apt to build up at an excessive rate; if capital is turned over too slowly, funds become stagnant and profitability suffers. The importance of capital turnover is vividly illustrated by comparing the profitability of the grocery chain and the chemical producer in the example. While the chemical firm achieved a profit margin more than five times as high as the grocery chain, its rate of capital turnover was less than a fifth that of the grocery. Since the return on net worth is equal to the profit margin multiplied by the capital turnover rate, both attained the same 12.5 percent return on their net worth.

SALES TO WORKING CAPITAL. The rate of working capital turnover can highlight a financial problem if it is either very low or very high. If the sales/working capital ratio is high, the business may owe too much, relying on credit as a substitute for an adequate margin of current operating funds.

SALES TO FIXED ASSETS. This ratio is less significant in itself than when compared with the same ratio for previous years. Such a comparison shows whether or not the funds used to increase productive capacity are being spent wisely. If comparable sales increases have failed to accompany sizable investments in fixed assets, then poor asset utilization is indicated.

CURRENT ASSETS TO CURRENT LIABILITIES. This ratio—also known as the "current ratio"—is one of the most commonly used ratios in balance sheet analysis. It gives an indication of the margin of protection for short-term creditors, a 2:1 ratio being considered a standard. In general, the more liquid the current assets, the less margin needed to cover current liabilities.

CASH TO CURRENT LIABILITIES. The ratio of cash and equivalent (marketable securities) to current liabilities—sometimes referred to as the "liquidity ratio"—is a supplement to the current ratio, indicating the firm's immediate ability to meet current obligations.

FIXED ASSETS TO NET WORTH. A firm's tendency to overinvest in fixed assets can often be identified by this ratio. A high ratio of fixed assets to net worth results in heavy depreciation and interest burdens, which can lead to serious profit problems should any sales difficulties be encountered.

CURRENT LIABILITIES TO NET WORTH. The ratio of current liabilities to net worth provides a means of evaluating a company's financial condition by comparing what is owed with what is owned. A high ratio—any value over 0.80—indicates that the firm is overly dependent on its creditors.

TOTAL LIABILITIES TO NET WORTH. Whenever the total liabilities/net worth ratio exceeds 1.00, it indicates that the creditors have a greater equity in the firm's assets than do the owners. Such top-heavy liabilities make the business extremely vulnerable to any unexpected contingencies and severely restrict the management's flexibility.

In summary, the final measure of management effectiveness can be found in a firm's balance sheet and profit-and-loss statements; there financial and operating strengths and weaknesses can be objectively identified and analyzed.

THE PROFIT/VOLUME (P/V) RATIO

In analyzing profits basically the same techniques are employed as in breakeven analysis, which is in fact simply a special case of profit analysis in which there is zero profit. Just as in breakeven analysis, graphical presentation of the data is most useful and easily understood.

The relationship between profit and volume is sought in the typical profit analysis study, volume being the independent variable and profit the dependent variable. Very few data are required to construct a simple P/V chart. Take, for example, the following condensed operating statement for a company.

Total sales		$1,000,000
Cost of operations		
Variable costs	$670,000	
Fixed costs	$230,000	
Total costs		$ 900,000
Net profit before income tax		$ 100,000

These data can be handled in generally the same way as in a conventional breakeven analysis. Figure 30 shows a P/V chart developed from this information.

In the P/V chart the vertical axis is divided into two parts, the top part representing net profit and the lower segment representing net loss.

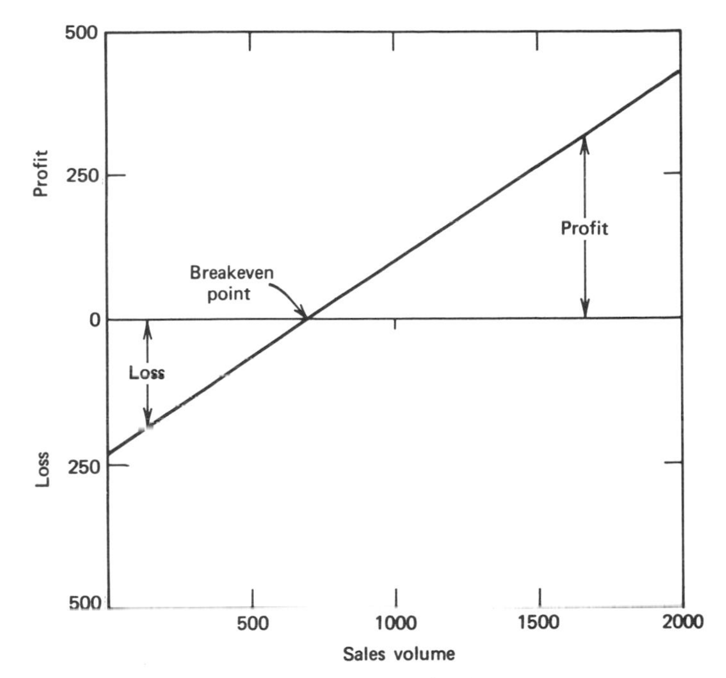

Fig. 30. *Typical P/V chart.*

The horizontal axis is drawn through the zero profit point on the vertical axis, representing the breakeven point. Units along the horizontal axis measure the company's total sales volume.

This P/V chart indicates the net profit or loss associated with any given level of sales.

At zero sales the net loss is equal to the amount of the overhead, or fixed, expenses. The P/V line, then, intersects the vertical axis at a net loss of $230,000 to be incurred at zero sales volume.

Since, according to the operating statement, the company's net profit at a sales volume of $1,000,000 was $100,000, the $100,000 figure can be plotted above the $1,000,000 point on the horizontal axis.

A straight line connecting these two points presumably gives the amount of profit or loss corresponding to any chosen level of sales, assuming that all other factors—such as the markups on direct costs, the amount of variable cost per dollar of sales, and the amount of overhead expense—remain the same. The point at which the P/V line intersects the horizontal axis—at $697,000 in this case—is the breakeven point. The area enclosed by the P/V line and the two axes to the left of the breakeven point represent net losses;

and the area bounded by the P/V line and the axes to the right of the breakeven point describe the firm's profits.

Other things equal, then, profits can be easily interpolated or extrapolated for any desired or expected sales volume. For this operation the $1,000,000 sales brought in $100,000 in profits. A sales volume of $1,500,000 should therefore yield $265,000 in profits, and a further increase in sales to $2,000,000 will result in profits of $430,000. A reduction in sales to the $500,000 level will result in a net loss of $65,000.

It is apparent from this P/V chart that, when the relationship between profit and volume is linear, a uniform change in profits is associated with a uniform change in volume. In this case each $100,000 change in volume causes a corresponding $33,000 change in profits.

APPLICATIONS OF THE P/V RATIO

The P/V ratio affords a convenient way to measure the firm's profit margin—the difference between total sales and total direct costs.

In the preceding example each $100,000 in sales was accompanied by a marginal profit of $33,000. The P/V ratio, defined as the ratio of incremental profit to incremental sales, is therefore 33.0 percent, or 0.33. This ratio can be used to compute the firm's breakeven point and net profits.

FINDING THE BREAKEVEN POINT. The breakeven point can be found by dividing total fixed cost by the P/V ratio. In the example

$$\text{breakeven point} = \$230{,}000/0.33 = \$697{,}000$$

CALCULATING NET PROFIT. The profit of any given sales volume can be found by multiplying the sales by the P/V ratio and deducting fixed expenses:

$$\text{net profit at } \$1{,}200{,}000 \text{ sales} = \$1{,}200{,}000 \times 0.33) - 230{,}000$$
$$= 396{,}000 - 230{,}000 = \$166{,}000$$

FACTORS AFFECTING PROFITS

As long as the relationships between overhead expenses, direct costs, and sales volumes remain constant, the preceding applications of the P/V ratio are valid. However, the entire profit situation changes whenever the markup, and consequently the variable cost rate, is changed.

The following formula is useful in identifying the impact of changes in fixed costs, markups, and sales volumes on a firm's profits.

$$P = S\left(\frac{M}{100 + M}\right) - F$$

where P is the net profit at sales volume S, M is the average percentage markup on direct costs, and F is the total amount of overhead or fixed expenses.

In referring again to the same example, the average markup can be found by dividing the total sales dollars by the total direct costs: 1,000,000/670,000 = 1.493, indicating an average markup of 49.3 percent. The net profit resulting from using this markup at the $1,000,000 sales level, then is computed as follows:

$$P = 1{,}000{,}000 \times \left(\frac{49.3}{100 + 49.3}\right) - 230{,}000$$

$$= (1{,}000{,}000 \times 0.33) - 230{,}000 = \$100{,}000$$

Changing the percentage markup on direct costs to 60 percent while keeping the other factors constant gives a profit of

$$P = 1{,}000{,}000 \times \left(\frac{60.0}{100 + 60.0}\right) - 230{,}000$$

$$= (1{,}000{,}000 \times 0.375) - 230{,}000 = \$145{,}000$$

In this example increasing the percentage markup from 49.3 to 60.0 percent increases the profit from $100,000 to $145,000. Thus a 20-percent increase in the markup results in a 45-percent increase in profits at these levels.

A change in the amount of overhead expense of course brings about a corresponding change in profits at any prescribed level of sales and markups; each dollar decrease in overhead raises profits by a dollar.

THE LAWS OF PROFITS

The relationships between profits, sales, overhead expenses, direct costs, and percentage markups lead to some general observations and conclusions. Briefly, these "laws of profits" are as follows.

- *All* costs—not just out-of-pocket direct costs—must be included when analyzing a firm's profit picture.
- Any change in fixed costs changes the net profit by a like amount, in the opposite direction.
- A given percentage change in fixed costs changes the firm's breakeven point by the same percentage and in the same direction providing other cost relationships hold steady.
- A change in the rate at which variable costs are incurred changes both the breakeven point and the net profit earned at a specified sales volume, if fixed costs remain constant. Increasing the variable expense rate

raises the breakeven point and lowers profits; decreasing the rate lowers the breakeven point and increases profits.

- An increase in the percentage markup applied to direct costs lowers the breakeven point and increases the profits associated with any sales volume. Decreasing the markup has the opposite effect.
- A change in both fixed costs and the variable cost rate has a significant effect on profits if they both change in the same direction; there is less impact if they move in opposite directions.

EFFECT OF OUTSIDE FINANCING ON PROFITS

In engineering projects the method of financing is often ignored in the feasibility study or preinvestment analysis. When the DCF approach is taken, it is simply assumed that, if the necessary capital can be obtained at a rate less than the indicated ROI, the project can be undertaken at a profit. However, if the project's expected ROI is less than the company's cost of money, the project should be avoided.

Obviously, no amount of outside financing can turn a basically unprofitable venture into a profitable one. But when the capital necessary to undertake a new venture can be obtained from outside sources at a rate less than what the project will return during its economic life, the ROI can be substantially amplified. Perhaps the most important single financial tool available to investors and businessmen is the leverage afforded by using other people's money (OPM). If, for example, a $100,000 investment were available that would yield a 15-percent annual return, and if the investor were able to borrow $75,000 at a 9-percent interest rate, his return on the $25,000 invested from his own pocket would be $(0.15 \times 100{,}000) - (0.09 \times 75{,}000) = \$8250/$ year. This amounts to a percentage return of $8250/25{,}000 = 33.0$ percent. In this case borrowing 75 percent of the required capital provides leverage which more than doubles the percentage ROI. If this same investor had $100,000 of his own money, he could seek out three more investments under comparable terms and end up with an annual return of $33,000—compared with only $15,000 which he would have realized had he attempted to finance the entire investment without outside sources of funds.

Leverage works in the other direction too, though, should the project fail to earn at the expected rate. When losses are incurred in a highly leveraged position, the effect can be devastating, and even earning just a little less than the cost of the borrowed money can have a damaging impact on a company's financial strength.

In using the same example as before, if the investor who borrowed 75 percent of his capital at a 9-percent interest rate managed to earn only 6 percent on the total amount invested in the project, he would end up with an out-of-pocket loss of $750; his project would earn $6000 while paying out $6750 in interest charges. If the investor had a more highly leveraged position —say in the 80–90-percent range—the impact would be far greater.

Figure 31 shows the effect of different levels of borrowing in this case, with all other conditions remaining the same. As indicated in the figure, the ROI goes from 15 percent when no outside borrowing is used, to 21 percent when half the total capital is borrowed, and up to 69 percent when 90 percent of the money is obtained by borrowing. Conversely, should the project earn only 6 percent on total capital, the percent return on the owner's equity would be accelerated in a negative direction.

MATHEMATICAL RELATIONSHIPS. The change in an investment's percentage ROI brought about by the leverage of borrowed money can be approximated

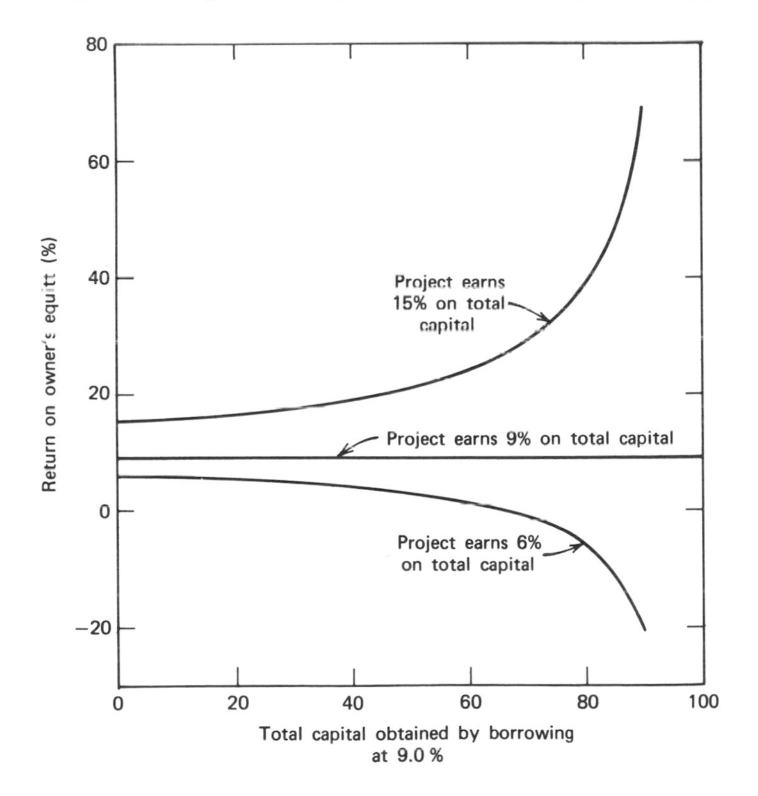

Fig. 31. *Example of effect of OPM on investment return.*

by using the following formula:

$$R_p = \frac{R_0 - pI}{1 - p}$$

where R_p is the return on investment when the portion p (expressed as a decimal) of the required capital is borrowed at a specified interest rate I. R_0 is the ROI that would be earned if all money were supplied internally, without borrowing.

Thus if a project were expected to earn a net return of 15 percent annually without benefit of outside financing, but if 80-percent financing could be arranged at a 9-percent interest rate, the new ROI for the project would be:

$$R_{0 \cdot 80} = \frac{15 - 0.80 \times 9.0}{1 - 0.80} = \frac{7.8}{0.2}$$

$$= 39 \text{ percent}$$

This formula is approximate only because it does not consider the timing of principal and interest repayments on the borrowed money, nor does it take into account the effect of interest payments on income taxes and, consequently, on cash flow patterns. It does, however, give a good, quick indication of the financial benefits made possible by using OPM.

Even in some instances in which the cost of borrowed money is less than the indicated ROI for the project, an investment still might be justified if there were no other way to go ahead with it and if the ROI were still adequate. A project starting with an ROI of 10 percent and financed one-half through borrowed 12-percent money still yields:

$$\frac{10 - (0.5 \times 12)}{1.0 - 0.5} = \frac{4.0}{0.5} = 8 \text{ percent}$$

In this situation, even though the project's ROI is reduced by the necessity to use OPM, the return might still be high enough to warrant consideration. If the only way that the project could be undertaken were by borrowing money, then borrowing under even these conditions might be acceptable.

REPAYMENT OF BORROWED MONEY

The timing of borrowed money, and especially the timing of principal repayments, is also a critical factor in leveraging an investment's ROI. This is especially noticeable in real estate transactions, in which property may still carry a substantial mortgage when it is sold.

For example, consider a tract of land purchased for $40,000, of which $10,000 is paid down and the $30,000 balance spread in equal payments over a 5-year period at 8-percent interest. The annual repayment schedule is:

Year	Principal	Interest	Total
0	10,000	0	10,000
1	6,000	2,400	8,400
2	6,000	1,920	7,920
3	6,000	1,440	7,440
4	6,000	960	6,960
5	6,000	480	6,480
Total	40,000	7,200	47,200

If the property were sold for a cash price of $45,000 at the end of the first year, the total payments to date would consist of $16,000 in principal and $2400 in interest. Part of the $45,000 received for the land would be used to pay off the $24,000 balance, leaving $21,000. The net capital gain on the property would be $5000, or 31 percent of the $16,000 actually invested. This gain is brought about by an increase in land value of just 12.5 percent coupled with the leverage of OPM.

If the property had been sold just before the first annual payment became due, the $45,000 would have paid off the $30,000 balance and the $2400 interest, leaving $12,600 and a clear $2600 gain on the 1-year investment of $10,000. Again, $5000 would be the capital gain and the $2400 would be tax-deductible as interest paid.

From a profitability standpoint, as long as an investment earns more than the capital costs, the investor benefits from the leverage of OPM. Therefore, to attain the highest possible ROI, the principal repayments should be deferred as long as possible. Briefly summarizing: Borrow as much as possible as cheaply as possible for as long as possible and pay it back as late as possible.

There are basically four ways an investment can be financed and repaid: (1) cash, (2) equal principal payments, (3) equal annual payments, and (4) interest only, with the principal repaid in the final year.

CASH ON THE BARRELHEAD. The simplest and least profitable way to handle a new capital investment is to use your own money. Many businesses were once conducted on this basis, and a few still are. If large cash reserves are available and there is no better use for them, then this approach at least keeps them from sitting idle. Table 55 shows the cash flows associated with a cash investment; the proposed venture is expected to generate operating profits of $6000/year before taxes and $3000/year after taxes. The entire $10,000

Table 55 Cash on the Barrelhead

Year	Operating Profit	Interest	Net Profit before Taxes	Net Profit after Taxes	Principal Payments	Net Cash Flow
0	0	0	0	0	10,000	(10,000)
1	6,000	0	6,000	3,000	0	3,000
2	6,000	0	6,000	3,000	0	3,000
3	6,000	0	6,000	3,000	0	3,000
4	6,000	0	6,000	3,000	0	3,000
5	6,000	0	6,000	3,000	0	3,000
Total	30,000	0	30,000	15,000	10,000	5,000

investment is made initially, followed by 5 years of cash inflows. Since there are no interest payments, the net cash inflows are the same as the net profits after taxes. This investment, set up in this way, yields an ROI of 15 percent on a DCF basis.

EQUAL PRINCIPAL PAYMENTS. Intermediate-term business loans are often repaid in equal principal installments, plus accrued interest on the unpaid balance. This approach is easy to handle, interest calculations are simple, additional principal payments can be made without disturbing a preset repayment schedule, and the lender can see the loan balance being reduced at regular intervals. Large land purchases are often owner-financed in this way, with payments made either annually or semiannually. Table 56 shows an example of this approach, using the same proposed venture as in the previous example. Here, though, only $2000 is applied as a down payment at the beginning, the $8000 balance being spread out in $1600 payments annually over the 5-year period. Since interest charges are computed on the unpaid balance, interest costs decrease each year. The net cash flows—the net profit after taxes, less principal payments—increase each year. The ROI earned by financing the project this way jumps up to 51 percent.

EQUAL ANNUAL PAYMENTS. Mortgages on commercial and residential buildings are generally set up on an amortization schedule, where a uniform periodic (usually monthly) payment covers both principal and interest. The principal portion increases with each payment, while the interest portion decreases as the unpaid balance declines. An amortized loan is relatively desirable, in that the heavy interest payments are incurred during the early years and the bulk of principal payments are deferred until later. In most amortized mortgages the payment schedule must be strictly adhered to, or the balance paid off in a lump sum. Table 57 describes the same investment as before, repaid in equal annual amounts (principal plus interest = $2004). Here net cash inflows are highest during the early years, thereby bringing the ROI up to 56 percent.

INTEREST ONLY. Best of all from a profitability standpoint is the situation in which the entire principal repayment is delayed until the project's final year and is then paid off in a single lump sum. Until then, only interest payments are made. Corporate and municipal bonds are set up this way, paying interest annually or semiannually until they are retired at a specified future date. As is the case with an amortized loan, early retirement can usually be made only by paying off the entire debt balance in a single amount. Table 58 illustrates this type of repayment schedule, again with a $2000 down payment required at the beginning. The $8000 principal balance is carried until the final year when it is paid off in one lump. Meanwhile, interest is paid annually

Table 56 Equal Principal Payments

Year	Operating Profit	Interest	Net Profit before Taxes	Net Profit after Taxes	Principal Payments	Net Cash Flow
0	0	0	0	0	2,000	(2,000)
1	6,000	640	5,360	2,680	1,600	1,080
2	6,000	512	5,488	2,744	1,600	1,144
3	6,000	384	5,616	2,808	1,600	1,208
4	6,000	256	5,744	2,872	1,600	1,272
5	6,000	128	5,872	2,936	1,600	1,336
Total	30,000	1,920	28,080	14,040	10,000	4,040

Table 57 Equal Annual Payments

Year	Operating Profit	Interest	Net Profit before Taxes	Net Profit after Taxes	Principal Payments	Net Cash Flow
0	0	0	0	0	2,000	(2,000)
1	6,000	640	5,360	2,680	1,364	1,316
2	6,000	531	5,469	2,735	1,473	1,262
3	6,000	414	5,586	2,793	1,590	1,203
4	6,000	286	5,714	2,857	1,718	1,139
5	6,000	149	5,851	2,925	1,855	1,070
Total	30,000	2,020	27,980	13,990	10,000	3,990

Table 58 Interest Only; Principal in Final Year

Year	Operating Profit	Interest	Net Profit before Taxes	Net Profit after Taxes	Principal	Net Cash Flow
0	0	0	0	0	2,000	(2,000)
1	6,000	640	5,360	2,680	0	2,680
2	6,000	640	5,360	2,680	0	2,680
3	6,000	640	5,360	2,680	0	2,680
4	6,000	640	5,360	2,680	0	2,680
5	6,000	640	5,360	2,680	8,000	(5,320)
Total	30,000	3,200	26,800	13,400	10,000	3,400

(computed at 8 percent of the $8000 unpaid balance), and cash flows are uniformly high until the final principal payment is made. Here ROI is an impressive 120 percent/year.

In these four examples the only differences are in the way principal and interest payments affect cash flows, yet the resulting ROIs vary by a factor of 8. In each case higher interest payments, reflecting deferred principal payments, bring higher investment returns.

SUMMARY

Financial analysis provides the means by which a firm's financial performance can be objectively evaluated. Several important financial ratios—developed from balance sheet and operating statement data—can be used to compare a company's performance with others in the same field, or with itself over a period of time. The profit-sales and profit-net worth relationships are particularly significant in showing a company's competitive strength and economic efficiency. The profit/volume ratio is also useful in anticipating the changes in profits associated with changes in sales; this ratio can be used to calculate breakeven points and profits under varying conditions. Many factors affect a firm's profits and profitability: changes in fixed costs; the rate at which variable costs accrue; prices, as determined by percentage markups on direct or variable costs; the sources and costs of outside funds employed; and the timing and methods of repaying borrowed money. Considerable leverage can be gained by using outside financing whenever the cost of borrowed money is less than the proposed venture is expected to earn. The project's overall return in such a case can be further enhanced by delaying repayment of the principal amount for as long as possible.

13

Forecasts and Forecasting

The term forecast has several dictionary definitions:

- to plan ahead;
- to foresee; to calculate beforehand;
- a prophecy or estimate of a future happening or condition.

Forecasting, then, requires that conclusions be drawn regarding future events by applying some sort of predictive model or technique to presently known facts or data. If the facts are correct and the technique is valid, then the conclusions must be accurate.

For those involved in the economic evaluation of engineering projects, forecasts of three broad types of "future happenings" are particularly important: economic, technological, and business conditions.

Economic conditions refer to the overall climate of the nation's aggregate economy, without particular reference to any specific industry; the gross national product (GNP) is probably the most important single indicator of general economic conditions. *Technological conditions* encompass a variety of factors affecting the development of new products, processes, and markets. *Business conditions* take into account the specific factors of price, cost and volume which ultimately determine the profitability of a given industry, company, or venture. In all cases the first step is to collect and organize the appropriate data.

SOURCES OF DATA

The most common sources of economic data include government agencies, trade associations, and business publications. When no published data are available, special tabulations or surveys can be made and in some cases may be warranted.

Much useful economic data can be found in readily available publications. The following list of information sources is not meant to be all-inclusive, but a familiarity with the references included in this list will prove extremely useful to anyone concerned with economic statistics.

Directories

Statistics Sources. Gale research, Detroit, Mich.

Executive's Guide to Information Sources. Gale Research, Detroit, Mich.

Business Trends and Forecasting Information Sources. Gale Research, Detroit, Mich.

Encyclopedia of Associations. Gale Research, Detroit, Mich.

Thomas Register of American Manufacturers. Thomas, New York.

Poor's Directory of Executives. Standard & Poor, New York.

Million Dollar Directory. Dun & Bradstreet, New York.

General Economic Data

Statistical Abstract of the United States. U.S. Government Printing Office, Washington, D. C. (annual).

Economic Almanac. National Industrial Conference Board, New York (annual).

Survey of Current Business. U.S. Office of Business Economics, Washington, D. C. (monthly).

National Economic Projections. National Planning Association, Washington, D. C.

Economic Indicators. U.S. President's Council of Economic Advisors, Washington, D. C. (monthly).

Handbook of Basic Economic Statistics. Economic Statistics Bureau, Washington, D. C. (monthly).

Industry and Product Data

U.S. Industrial Outlook. U.S. Bureau of Defense Services Administration, Washington, D. C. (annual).

Predicasts. Predicasts, Cleveland (quarterly).

Census of Business. U.S. Bureau of the Census, Washington, D. C.

Census of Manufactures. U.S. Bureau of the Census, Washington, D. C.

Census of the Mineral Industries. U.S. Bureau of the Census, Washington, D. C.

Census of Population. U.S. Bureau of the Census, Washington, D. C.

Minerals Yearbook. U.S. Bureau of Mines, Washington, D. C. (annual).

Construction Review. U.S. Bureau of Defense Services Administration, Washington, D. C. (monthly).

Area Information

Commercial Atlas and Marketing Guide. Rand-McNally, Chicago.

Editor and Publisher Market Guide. Editor and Publisher, New York (annual).

Survey of Buying Power. Sales Management Magazine, New York (annual).

Municipal Yearbook. International City Manager Association, Chicago (annual).

Financial Data

Quarterly Financial Report for Manufacturing Corporations. U.S. Federal Trade Commission and U.S. Securities and Exchange Commission, Washington, D. C. (quarterly).

Leading U.S. Corporations. News Front Magazine, New York.

Moody's Handbook of Common Stocks. Moody's Investors Service, New York (quarterly).

Stock Guide. Standard & Poor, New York (monthly).

Value Line Investment Survey. Arnold Bernhardt, New York (weekly).

General Business and Financial Periodicals

Wall Street Journal. Dow Jones, New York (daily).

Business Week. McGraw-Hill, New York (weekly).

Forbes. Forbes, New York (bimonthly).

Fortune. Time, New York (monthly).

Dun's Review. Dun & Bradstreet, New York (monthly).

Obviously, not all these references will be pertinent to the engineer's interest, and other specialized publications may offer more direct benefits. But anyone involved in economic analyses should be aware of the overall economic environment—past, present, and future.

BASIC FORECASTING TECHNIQUES

Forecasting is said to be "sometimes a science, sometimes an art, and most often a little of both." Essentially an economist's tool, forecasting is a necessary part of planning most engineering projects.

The accuracy of economic forecasts can often mean the difference between business success and failure. Management decisions in the areas of production, financing, marketing, inventory planning, and capital investment depend to a large extent upon the information developed in economic forecasts.

Several assumptions are helpful in approaching any business forecast:

1. The magnitude of, and the relationships between, most economic factors change slowly.
2. What happens in the future is strongly influenced by what is going on at present.
3. Past experience offers the best guide to these cause-and-effect relationships.

In predicting the future, then, it is necessary to "know the past and understand the present." A forecast of economic activity in any business or industry involves an analysis of past trends, extrapolation of these trends into the future, and adjustment, refinement, or revision of the projected trends to allow for any factors expected to cause deviations from the established trends.

The specific techniques to be employed in economic forecasting depend largely upon how far into the future the forecast extends. For extremely short-range forecasts—perhaps covering only a week or a month—it may be safe to assume that the immediate future will be very much like the present, or at least that recent trends will continue for a while. Using this approach proves adequate most of the time, significant errors occurring only when there is a major reversal of trends.

A second method of forecasting, frequently used for intermediate-term forecasts, employs "leading" indicators to predict coming changes. Next year's construction activity, for example, will reflect this year's contract awards. Some of the most important indicators used by economists include interest rates, contract awards, stock prices, manufacturers' new orders, freight car loadings, indexes of industrial production, corporate profits, and business failures.

A similar procedure can be used effectively with "concurrent" indicators instead of leading indicators, providing the concurrent indicators can be projected with some degree of certainty. Construction volume, for example, can be correlated closely with the GNP. If a good estimate of next year's GNP is available, the total contract construction volume can be estimated within 1 or 2 percent. When the construction-GNP relationship is expressed mathematically, it becomes in effect a third type of forecasting technique.

This third type of economic forecast uses an econometric model—a mathematical-statistical description of an economy, business, or industry. This approach is especially useful in long-range forecasts, when there are many unknown factors involved. By expressing these factors in equations, they can be easily varied and the effects of changes on the ultimate outcome can be assessed. In this way many different assumptions can be used and their results examined; as a result, the most likely outcomes can be predicted, as well as the range of possible outcomes.

TREND AND REGRESSION ANALYSIS

Business economists employ numerous different techniques in making their forecasts, with widely varying levels of precision and sophistication.

There is scarcely a limit to how complicated a problem can be made, how far back data can be collected, or how much data can be included. But in many instances the *Principle of historical continuity* yields as reliable a prediction as the most sophisticated econometric models.

The Principle of Historical Continuity

When predicting future prices or weather, unless there is some specific reason to believe otherwise, the safest bet is usually to assume that tomorrow will be pretty much the same as today. This is known as the principle of historical continuity.

The principle of historical continuity is a direct outgrowth of the ancient "lost horse" technique, which states that the best way to find a lost horse is to go where it was last seen and start looking in whatever direction it was heading.

In fact, patterns and trends established in the past often prevail with sufficient frequency to warrant using them as a basis for predicting the future. Extending the already established trend into the future involves *extrapolation.* Identifying the appropriate trend, though, requires some careful thought.

To lend the quality of statistical objectivity to predictions made through extrapolation, a regression line can be fitted to the available data and then extended. Probably the best way to assure a valid fit of the regression line is to apply the principle of least squares.

THE METHOD OF LEAST SQUARES

In fitting a straight line to a series of points, there is bound to be some scattering of the points about the line. In order to find a regression line around which the degree of scattering is a minimum, the *method of least squares* is commonly used.

A regression line selected according to the principle of least squares is, by definition, the line that results in the smallest possible sum of the squared vertical deviations of the points from the line.

An equation for this least squares regression line can be developed mathematically. However, if all that is desired is an indication of the magnitude and direction of a trend, a quick graphical solution is usually adequate.

Figure 32 illustrates the graphical method of solution. Here there are seven points having equal horizontal spacing. It is desired to fit a least-squares regression line to these points.

Starting from the point on the extreme left, draw a straight line toward the second point, extending this line just two-thirds of the horizontal distance between the two points.

Then, continuing from this point at which the line is stopped, aim toward the third point and draw a line in that direction for the same horizontal distance—that is, two-thirds of the horizontal distance between each pair of adjacent points.

Repeat this procedure point by point until the "target" point is the last one, at the extreme right. The final line segment drawn will end exactly two-thirds of the horizontal distance from the extreme left-hand to the extreme right-hand point. This final point lies on the least-squares regression line.

To locate the second point needed to define the regression line, start with the point on the far right and work back toward the left, point by point, each

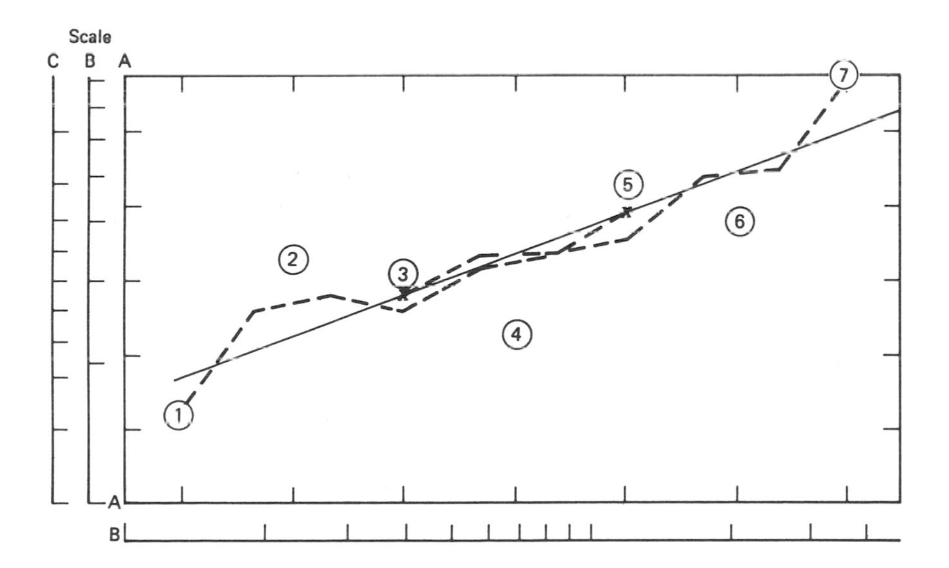

Fig. 32. *Graphical solution for least-squares regression line.*

time moving a distance equal to two-thirds of the horizontal distance between adjacent points.

The dotted lines in the figure show the course of the line developed from left to right, ending in an "x" just below the fifth point; and from right to left, ending at the "x" just below the third point.

A straight line passing through the two x's represents the least squares regression line—or the unique straight line that minimizes the sum of the squares of the vertical distances from each point to the regression line. The sum of the squared deviations is known statistically as the *variance*; its square root is known as the *standard deviation.*

Calibration of the scales makes no difference, so long as the points are uniformly spaced and measurements are made at a constant horizontal increment. Vertically, for example, scale A is arithmetic; scale B is logarithmic; and scale C represents a normal probability distribution.

On the horizontal scale, scale A is arithmetic and scale B is logarithmic. However, if the logarithmic scale is used, in order for the points to be horizontally equidistant, they must be plotted at increments of 2^N—at units of 1, 2, 4, 8 . . ., for example. An arithmetic scale is far more convenient.

APPLICATIONS OF TIME SERIES ANALYSIS

Time series analysis is one of the most useful applications of this graphical technique in economic analysis; the horizontal scale is ordinarily arithmetic, measuring years, and the vertical scale may be either arithmetic or logarithmic, measuring dollars, ratios, or index numbers and indicating either absolute changes (on an arithmetic scale) or rates of change (on a logarithmic scale).

When plotting numerical values by years, how closely the individual points fit the least-square regression line is a direct indication of stability, while the slope of the line measures the trend. Most mature public utilities' earnings, for example, plot nearly as a straight line, indicating low variance and high stability and thus offering a relatively high level of confidence in predicting future earnings.

Firms in less stable industries, especially those subject to rapid technological change, intense competition, or highly erratic market demands, are apt to show much greater variances, and projections of their future earnings are therefore subject to a much higher degree of uncertainty.

EXTRAPOLATION

The best guide to an individual's future behavior is often indicated by his past behavior in similar situations. Such is also the case with economic data;

the best estimates of the future can often be made by a careful study of the past. Unless there is some reason to believe otherwise, whatever trends have taken place in the past generally can be expected to continue into the future. This technique of economic forecasting, in which established trends are assumed to continue, is called extrapolation.

Extrapolation refers simply to the extension of a series of numbers beyond its original range by adding other numbers at either end of the series, which fit the established pattern of the original series. In other words, the values of points lying beyond a known interval are estimated solely on the basis of known values lying within the interval, without regard to any underlying causes.

Extrapolation is the simplest yet one of the most effective and useful forecasting techniques. While most commonly applied to time series data, it can also be used in other types of correlation analyses. Since the basic assumption in forecasting by extrapolation is that the future is a direct reflection of past trends, the basic problem is to identify the appropriate growth trend indicated by historical data. This is done in several steps:

1. Plot the pertinent data on graph paper.
2. Try out different curves to see which fit the data.
3. Select the curve that best describes the data.
4. Fit the appropriate trend line to past data, using some accepted statistical technique.
5. Extend the trend line into the future.
6. Read the expected future values off this extended trend line.
7. Make adjustments in these future values based on any other available knowledge.

There are numerous curves that can apply to various types of data, certain types of curves being especially typical of certain types of economic data. Only a few of the most commonly encountered trend curves are discussed here.

TYPES OF TREND CURVES

Generally, the most useful trend curves encountered in economic forecasting are the arithmetic trend, the semilogarithmic trend, the logarithmic trend, and the S trend.

These curves are illustrated in Figures 33–36, plotted on arithmetic scales to show their general shapes. They plot as straight lines on different types of graph paper.

THE ARITHMETIC TREND. The arithmetic relationship shown in Figure 33 plots simply as a straight line on conventional arithmetic graph paper. It

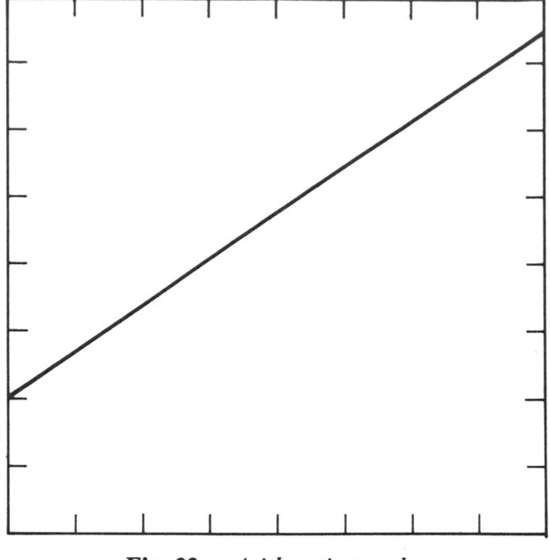

Fig. 33. *Arithmetic trend.*

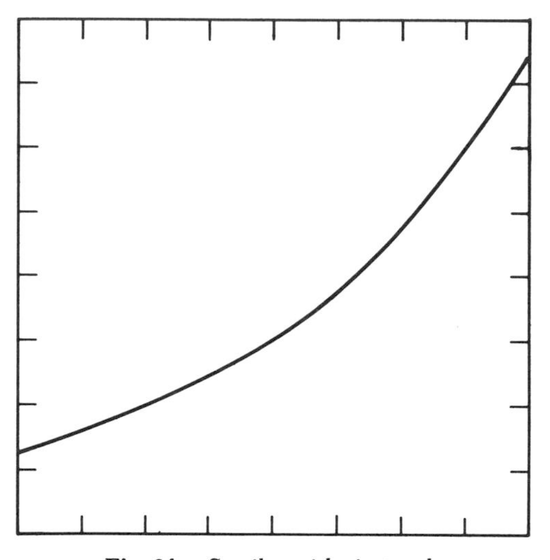

Fig. 34. *Semilogarithmic trend.*

represents, in a time series, a constant numerical increase per unit of time. The arithmetic curve has the general form $y = a + bx$.

THE SEMILOGARITHMIC TREND. The semilogarithmic trend (Figure 34) depicts a constant *rate* of change (as opposed to a constant numerical change in the arithmetic trend). Thus the change occurring during each time period is a

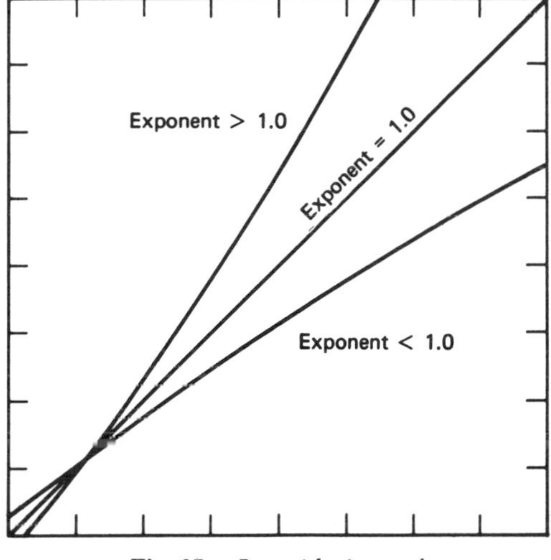

Fig. 35. *Logarithmic trend.*

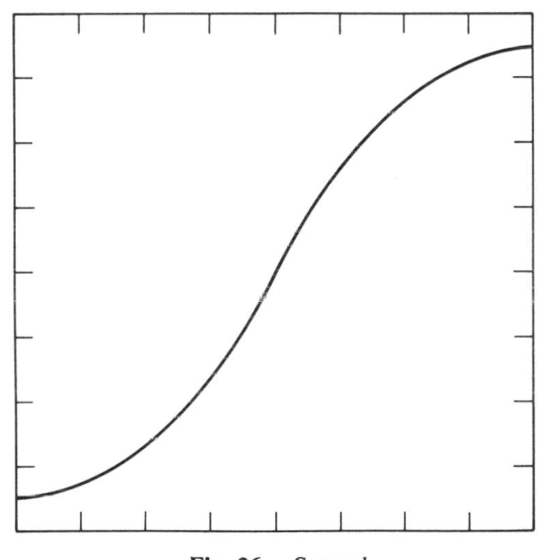

Fig. 36. *S trend.*

fixed percentage of the cumulative total at the end of the preceding time period. The semilog trend is typical of population growth and many other types of economic data, such as per capita consumption of utility services. Values following a semilog trend plot as a straight line on semilog graph paper, which is the most convenient way to extrapolate and present the data. This curve follows the form $y = ab^x$.

THE LOGARITHMIC TREND. The logarithmic trend takes an exponential form, as illustrated in Figure 35. If the exponent is greater than 1, it will slope upward; if the exponent is less than 1, the curve will slope downward; and if the exponent is exactly 1, the trend will in effect be arithmetic and will plot as a straight line either on log-log or arithmetic graph paper. Cost-capacity data typically follow the exponential form shown in Figure 35, plotting as a straight line on log-log paper and having exponents less than 1. Doubling a facility's capacity, for example, may increase the cost by only half, representing a logarithmic trend having an exponent of about 0.6. Mathematically, this curve can be defined as $y = ax^b$.

THE S TREND. The S-shaped curve shown in Figure 36 plots cumulatively a "normal" or binomial frequency distribution of data. This type of curve is especially valuable in forecasting future sales of a specific product, since the life-cycles of most products generally follow this shape. They typically start slow, pick up momentum as they gain market acceptance, and then taper off gradually as the market becomes saturated and newer products begin replacing them. A symmetrical S curve plots as a straight line on probability graph paper. To determine where a specific product stands on its growth curve, an estimate must first be made of its total potential cumulative sales; then its cumulative sales history to date (expressed as a percentage of the estimated potential) can be plotted on probability paper. If the resulting plot is a straight line, the estimate was good; otherwise, a new estimate of sales potential should be made and the entire procedure repeated until the plotted points come out on a straight line on probability paper. Extending the straight line gives both the number of years until the potential is reached and the annual sales increments from which the sales for any given year can be calculated. Other common types of S-shaped curves include the logistic and Gompertz curves, and many theoretical cumulative frequency distributions plot in the S shape.

TECHNOLOGICAL FORECASTING

Intelligent planning of engineering projects requires numerous assumptions about the future of technology, so that the impact of future changes on present designs can be anticipated. This requires prediction of what is by definition essentially unpredictable: innovation and technological change.

The only certainty about the future is that it will be different from the present, just as certainly as the present is different from the past. Determining just how much difference can be expected, and defining the ways in which the differences will be manifested, are the primary objectives of technological forecasting. Whether the engineer's interests are in water supply and wastewater disposal, electric power generation, transportation systems, environmental pollution, or other fields, technological forecasts can play an important part in decision making.

Technological change is a result of economic, social, and political conditions which largely determine the extent and direction of technological effort. Consequently, the first step in making a technological forecast is to analyze what has already taken place for technology to arrive at its present status. Then the future can be predicted, based on where society in general is likely to be headed. A concern with environmental quality, for example, is certain to have significant effects on technology related to wastewater treatment, solid waste disposal, air pollution control, and noise abatement programs.

The second step in the technological forecast is to predict both the qualitative and quantitative changes in the technology under consideration. In taking air pollution as an example, the quantity of particulate matter discharged into the atmosphere by a particular industry in the year 2000 will be a function of at least three independent factors: (1) changes in the amount of pollutants produced, brought about by changes in production levels and processes; (2) changes in the number or proportion of plants in that industry that have installed particulate control devices; (3) the collection efficiency of the control systems in operation. Production levels depend chiefly upon economic considerations; pollution control practices are most influenced by legal restraints in the form of politically established emission control standards; and production processes and equipment efficiency are engineering functions.

The third step, then, is to reconcile the results of the second-step forecast with the requirements defined in the first step. In the preceding example, if emission levels are still higher than society will accept, then the original forecasts must be modified. Either technology must conform to society's demands, or society must adapt itself to technology. The forecaster must somehow end up with a set of forecasts—economic, sociological, and technological—that are compatible with each other.

FORECASTING TECHNOLOGICAL EFFICIENCY

The efficiency of almost everything increases with time (*time*, not *age*). A 1970 generating plant is more efficient than a 1940 plant was when it was new; and a generating plant built in the year 2000 will be more efficient than the plants installed in 1970.

The following universal technological efficiency forecasting technique is offered for use in situations in which no better data are available. The curve in Figure 37 represents the improvement in technological efficiency that is achieved with time. The vertical scale measures efficiency, while the hori-

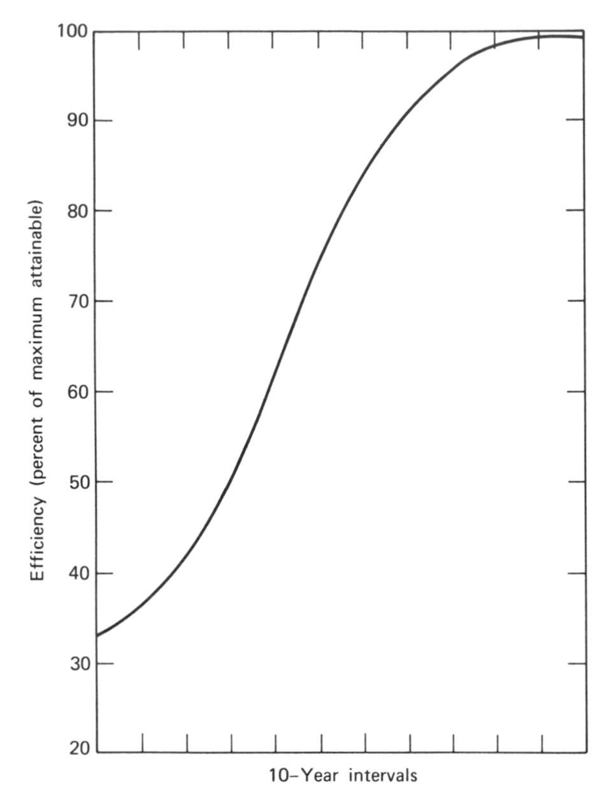

Fig. 37. *Pattern of technological improvement.*

zontal scale represents the time interval required to achieve, in practice, any desired increase in efficiency.

The curve indicates, for example, that if something is currently operating at 50-percent efficiency, its efficiency should rise to 62 percent in 10 years. Things presently at a 75-percent efficiency level are expected to increase to 84 percent in 10 years; but if the efficiency is already 95 percent, the expected 10-year improvement will raise it only to 98.5 percent. Nothing is ever 100 percent efficient.

This curve does not necessarily imply that a specific product will be more efficient in the future than it is now; instead, the curve is based on the assumption that more efficient ways will be found of performing the same function.

Thus it does not mean that the efficiency of mechanical dust collectors will increase, say, from 80 percent to 88 percent over the next decade. It does indicate that, if the average efficiency of particulate control systems currently being installed on industrial plants is 80 percent, the efficiency of systems installed in 1980 will average 88 percent. The increase in system efficiency may be partially due to increased cyclone efficiency, but will probably be more strongly influenced by a shift toward electrostatic precipitators, fabric filters, and other types of more efficient equipment.

Similarly, near-term increases in the efficiency of wastewater treatment plants will be a result of more widespread use of presently available technology rather than any innovative treatment concepts. To raise the efficiency of sewage treatment, activated sludge systems will prevail over trickling filters to an increasing degree, new treatment processes will be developed, and provisions for tertiary treatment will become more prevalent.

It must be realized that technological change does not necessarily represent technological improvement, and what technology is capable of doing is not necessarily what technology does. Nevertheless, the technological forecast can provide the engineer with much information useful in long-range planning and help him make intelligent decisions in view of an always uncertain future.

FORECASTING BUSINESS SUCCESS

Most new business ventures fail outright, while many others never yield the returns that were expected for them. Yet the great majority of these business failures can be attributed to circumstances or conditions that were known prior to undertaking the project, or to factors that could easily have been identified well in advance of making any major financial commitments. For some reason, whether enthusiasm or ignorance, management simply failed to recognize the importance or existence of these factors.

To be financially successful, a new product or new service venture must have three broad bases of support:

1. A superior product.
2. A capable company.
3. A favorable environment.

The whole problem of forecasting business success in a particular venture, whether it involves a new product or a new service, can be visualized conceptually as a three-legged milking stool, the product, company, and environment representing the legs supporting the overall venture. For the venture to be stable, its supports must be strong and balanced. A single short or weak leg can cause the entire venture to topple, or to at least become uncomfortably shaky.

While the specific problems encountered in different types of business activity vary widely and the relative importance of factors are not the same in each case, the primary determinants of success are very much alike; *what* is

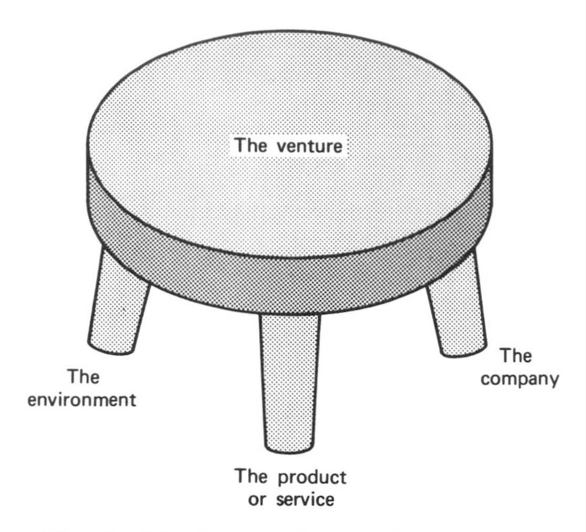

Fig. 38. *The elements of success for new ventures.*

being sold, *who* is attempting to sell it, and *where* it is sold. In other words, the *product*, the *company*, and the *market* together determine how successful the venture will be.

The *product* or service must be objectively evaluated in terms of its competitive performance, salability, and defensibility. Each of these categories is further divided into several individual factors, such as effectiveness, reliability, simplicity, convenience, appearance, and many others, to determine how favorably the proposed product compares with the products against which it must compete in the marketplace.

The *company's* capabilities must include three broad areas: marketing, technology, and production, marketing probably being the most important of the three. Again, each of these three areas of capability can be broken down into many distinct factors which can be evaluated individually.

The *environment* refers to external considerations such as the market, competition, suppliers, and government. The company must react to environmental conditions but can seldom control them; it must sell its products in the existing environment, subject to numerous conditions and constraints imposed by competition, legislation, and the general economy. As before, the pertinent factors can be readily identified and evaluated for a specific venture.

Finally, the overall *venture's* success is measured in terms of its investment returns, strategic benefits, and other outcomes determined by a combination of the product, company, and environmental factors.

By putting together all the factors relating to product success, it is possible to predict, in advance and with fair reliability, the probability of the product's ultimate success in the marketplace. Table 59 gives a rough indication of the relative probability of success for new product ventures, based only on whether the proposed new product is better than, equal to, or not as good as competition with respect to the three broad rating factors—the product, the company, and the environment.

The assumption is made in this table that all three factors are of equal importance. A venture rating high in all three categories is assigned a relative probability of success of 100, and all other combinations are related to this index value.

The techniques by which several hundred individual factors can be incorporated in the venture evaluation model can become quite sophisticated and far more useful and meaningful than the ultrasimple method summarized in the table. For a more complete discussion of this whole approach to venture evaluation, the reader is referred to: Maillie J. B., *Venture Evaluation Rating System*, Midwest Research Institute, Kansas City, Missouri.

Even a strong company cannot successfully sustain a market for a poor product, nor can a company with inadequate capabilities in marketing be successful even with a superior product. It is possible for a first-rate company to introduce a product that is far superior to all other available products and then find that there is no consumer demand for the product regardless of its superior attributes. Only if a strong, capable company offers a superior product in a well-defined market in which a strong demand exists is there a good chance of success.

There are of course exceptions. Whenever a new company is formed, it faces a handicap—similar to a pinch hitter going into a baseball game with a one-strike count. A new company with a good product in an uncertain market has two strikes against it at the beginning, as would an established company with an inferior product in a weak market, or a company with limited marketing capabilities and an inferior product in a strong market.

Still, there have been instances of new companies offering new products in untested markets and being successful, especially in low-priced "fad"-type consumer items. There are, however, many more examples of such ventures that have failed miserably.

It is possible to take all the hundreds of individual factors affecting the success of a new product, service, or venture, put them together in a structured, well-ordered model, and draw some meaningful conclusions from them. Adjustments can be made for the type of product or industry; uncertainties in judging certain factors can be considered; and operational differences in companies and unusual conditions can be anticipated and incorporated in the model. The results of employing an objective, analytical ap-

Table 59 The Probability of Success for New Product Ventures

Rating Factors*			
The Product	The Company	The Environment	Relative Probability of Venture Success
+	+	+	100
+	+	=	60
+	+	–	25
+	=	+	60
+	=	=	40
+	=	–	15
+	–	+	25
+	–	=	15
+	–	–	5
=	+	+	60
=	+	=	40
=	+	–	15
=	=	+	40
=	=	=	25
=	=	–	10
=	–	+	15
=	–	=	10
=	–	–	3
–	+	+	25
–	+	=	15
–	+	–	5
–	=	+	15
–	=	=	10
–	=	–	3
–	–	+	5
–	–	=	3
–	–	–	1

*+ Represents good, or better than competition; = represents average, or comparable to competition; – represents poor, or inferior to competition.

proach such as this to venture evaluation have, in practice, repeatedly shown several important advantages:

- It provides a convenient, economical, and versatile method for quickly and objectively screening large numbers of proposed new ventures.
- It compares the relative merits of proposed ventures on a consistent and well-defined basis, both with each other and with existing product lines or services.
- It directs management attention to all factors affecting venture success, thus preventing enthusiasm for a few favorable points from carrying a proposal through.
- It effectively concentrates and utilizes the collective observation, experience, and judgment of company management.
- It identifies and measures present and potential strengths and weaknesses.
- It reduces the chances of venture failure by pinpointing the potential causes of failure for a specific venture at the beginning.
- It maintains continuity and consistency in a firm's overall venture evaluation program.

SUMMARY

Forecasting involves making predictions of future events or conditions based on present knowledge. Forecasting is especially important and useful to the engineer working in economic, technological, and business areas, and is accomplished by applying an appropriate forecasting technique to available, reliable data. The forecasting technique generally involves analysis of past trends, extrapolation of these trends into the future, and adjustment or refinement of the resulting projections. Many useful economic data can be obtained from readily available published sources. Once the necessary data have been accumulated, basic statistical techniques can be applied for trend and regression analysis, time series analysis, and extrapolation. Technological forecasting, an important part of many engineering proposals, requires analysis of the impact of economic, social, and political considerations as they influence technological development and change. Success in a new business venture is dependent on the venture's having three broad bases of support: (1) a superior product or service, (2) a capable company, and (3) a favorable environment in which to operate. Being deficient in any of these three areas substantially reduces the chances for success in the proposed venture, and a careful evaluation of all pertinent factors is therefore essential in considering the potential attractiveness of a new venture.

14

Probability, Risk, and Uncertainty

Most management decisions are based on beliefs—or on judgments of probabilities—rather than on certainties, with experience usually providing the basis for these judgments. The application of statistical techniques and probability theory can do much to help define uncertain events in objective, measurable terms.

The usefulness of statistics as a tool for engineering-economic analysis depends to a large extent upon the availability of suitable data, since the inferences to be drawn from statistical analysis can be no better than the raw data used as a basis for the study. Typical data suitable for statistical treatment include information on sales, estimated and actual costs, and profits.

In accumulating data for economic analysis and for eventual use in business planning, the data must be arranged and grouped in some easily interpreted fashion. Usually, this requires compilation of frequency distributions and construction of a frequency distribution curve.

THE FREQUENCY DISTRIBUTION

A frequency distribution is simply a grouping of statistical data after the data have been sorted and arranged in logical order, such as from the smallest to the largest. The purpose of the frequency distribution is to show the frequency of occurrence of data within each group. The frequency distribution is particularly valuable in analyzing relationships involving large quantities of data that might otherwise be difficult to interpret.

Often simple arithmetic averages are of only limited value. With the frequency distribution, in addition to identifying the average values of a mass of data, the entire range of values can be clearly defined.

In constructing a frequency distribution curve, raw data are grouped into their appropriate classifications by dividing the overall range of values covered by the data into convenient-sized groups and then tallying the data falling within each group. Table 60 illustrates some grouped data used in analyzing the accuracy of engineers' estimates.

Table 60 Accuracy of Engineers' Estimates

Ratio, Actual Cost to Estimated Cost	Percent of Projects	Percent Costing More Than
0.75 - 0.80	3.0	97.0
0.80 - 0.85	15.2	81.8
0.85 0.90	19.7	62.1
0.90 - 0.95	10.6	51.5
0.95 - 1.00	15.2	36.3
1.00 - 1.05	15.2	21.1
1.05 - 1.10	10.6	10.5
1.10 - 1.15	4.5	6.0
1.15 - 1.20	3.0	3.0
1.20 - 1.25	3.0	0.0
	100.0	

In Table 60, actual costs for a large number of jobs were converted to ratios, related to the engineers' estimated costs on the same projects. The ratios were then grouped into 10 narrower ranges, each range representing an interval of 5 percent of the estimated cost. This group of data has a mean value of 0.96, although the individual jobs have actually cost from 25 percent more to 25 percent less than estimated. More than half the jobs fell within ± 10 percent of their estimated costs, and 30.4 percent were within 5 percent of the amount estimated.

Figure 39 shows the frequency distribution in graphical form as it is plotted from the data in Table 60. This figure is generally typical of frequency distribution curves in that beginning from zero the curve rises to a maximum somewhere around its average value (in this case the highest frequency of occurrence is in the 0.85–0.90 interval), then tapers off, and finally reaches zero again. If the curve were perfectly symmetrical, it would usually be referred to as a normal curve representing a normal distribution, and the highest point on the curve would correspond to the arithmetic average of all the data. In statistical terms the distribution of Figure 60 is "positively skewed," signifying that the distribution tails off toward the right at a slower

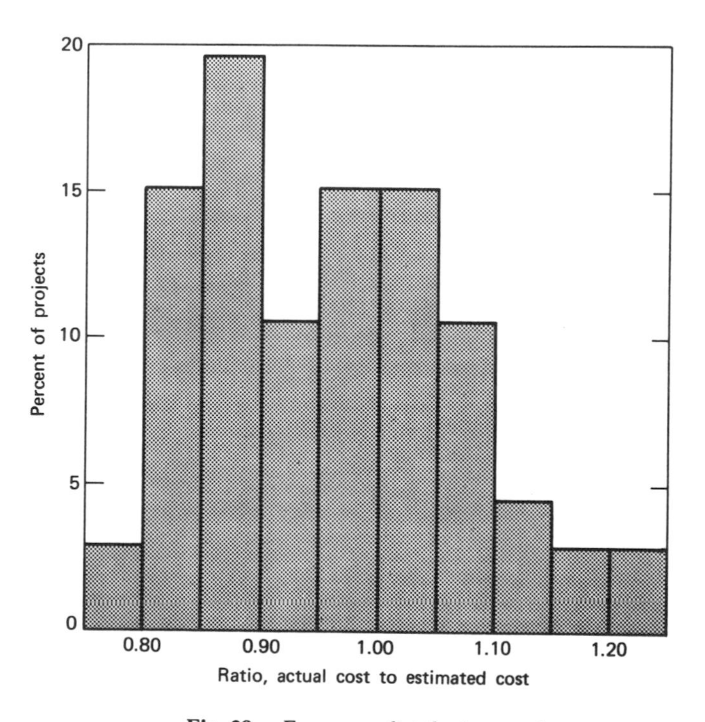

Fig. 39. *Frequency distribution graph.*

rate than toward the left, and meaning that the range of values on the high side of the average value is greater than the range of values on the low side.

THE CUMULATIVE FREQUENCY DISTRIBUTION

The right-hand column in Table 60 represents the total number of jobs having actual/estimated cost ratios equal to or greater than the range given in the left-hand column. This total is found by adding together the total number of estimates falling within each interval, starting from the end and working backward. For example, 51.5 percent of the jobs had an actual/estimated cost ratio of 0.95 or higher, while just 10.5 percent fell above the 1.05–1.10 interval.

The cumulative total number of jobs falling above each range of values is plotted in Figure 40. The frequency distribution data, plotted cumulatively, start at 100 percent of the total number of jobs included in the analysis. From this point the curve decreases rapidly until it reaches approximately the middle, at which time the rate of decline decreases and the curve tapers off gradually until reaching zero.

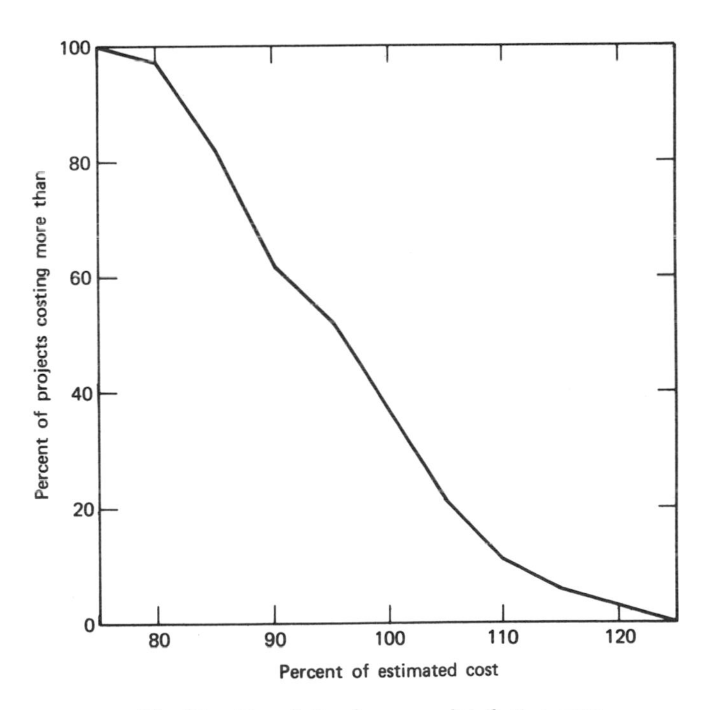

Fig. 40. *Cumulative frequency distribution curve.*

The cumulative frequency distribution curve can be used in predicting the probability of an event's occurrence, assuming that historical data are indicative of the future. For example, experience has shown in this case that 36.3 percent of the jobs cost more than the engineers' estimates. By adding a 5-percent "contingency" allowance to the estimated cost, the percentage of jobs costing more than estimated drops to 21.1 percent; a 10-percent contingency allowance covers all but 10.5 percent of the jobs; and a 15-percent allowance leaves only 6.0 percent of the jobs exceeding the estimate. While estimating accuracy is not improved by inclusion of this contingency allowance, the probability of exceeding the estimate is reduced substantially and the practice may be worthwhile from the standpoint of capital budgeting.

PROBABILITY AND EXPECTATION

Probability refers to the frequency of occurrence of an event, measured as the ratio of the number of different ways that the specified event can happen to the total number of possible outcomes.

Probabilities are expressed as a fraction between 0.0 and 1.0, or as a percentage from 0 to 100. A probability of 0 denotes an event that cannot happen, while a probability of 1.0 or 100 percent represents absolute certainty. This numerical description of the likelihood of an event's occurrence makes possible an objective measure of many situations that could otherwise be expressed only instinctively or intuitively.

The most important principles of probability can be summarized in six general rules.

RULE 1. Probabilities are, by definition, always between 0 and 1.0 (or 0 and 100 percent).

RULE 2. The probability that an event will occur, plus the probability that it will not occur, equals 1.0 (or 100 percent).

RULE 3. The probability that one or the other of two mutually exclusive events (events that cannot happen at the same time) will occur is equal to the sum of their individual probabilities of occurrence.

RULE 4. The probability that two or more independent events will happen simultaneously is equal to the product of their individual probabilities.

RULE 5. The probability that a specified event will occur when its occurrence is contingent on another event taking place first is equal to the probability of the first event times the probability of the second event's occurring after the first has already happened.

RULE 6. The probability of having either one or the other of two events happen when the events are not mutually exclusive is equal to the sum of their individual probabilities, less the probability of both happening at the same time.

Mathematical expectation provides a means of striking a balance between the probability of an event's occurrence and the benefits to be gained by its occurrence.

Expectation is calculated by multiplying the amount to be gained by a particular outcome by the probability of that outcome's occurrence. As such, mathematical expectation represents the average benefits that can be expected to result from a given situation over a long period of time, taking into account both the gains from successes and the losses from failures.

APPLICATION OF PROBABILITY TO A PROJECT

By incorporating probability factors in the

$$\text{profit} = (\text{unit price} - \text{unit cost}) \times \text{volume}$$

relationship, the entire range of possible outcomes resulting from variations in the individual factors can be identified and a project's profit expectations can be determined. Table 61 summarizes a simple example.

Here a product is to be manufactured. Its price will be determined by competitive factors and will range from \$40 to \$60 per unit with the probabilities shown in the table. Similarly, unit costs of either \$30, \$40, or \$50 may be incurred at the probabilities shown, and sales volume may vary according to some outside, uncontrollable influences. In this example, then, the three independent variables—price, cost, and volume—are totally independent of each other.

The probability of occurrence of any given combination of the three independent variables (from the fourth rule of probability) is equal to the product of their individual probabilities. Since there are three different values for

Table 61 Probable Values for Price, Cost, and Volume in a Manufacturing Operation

Unit Price		Unit Cost		Sales Volume	
$	Probability	$	Probability	Units	Probability
40	0.2	30	0.3	1000	0.3
50	0.5	40	0.6	2000	0.4
60	0.3	50	0.1	3000	0.3

each of the three variables, a total of $3 \times 3 \times 3 = 27$ different outcomes are possible. These 27 outcomes are summarized in Table 62.

The outcomes range from the worst possible combination of the three variables (low unit price, high unit cost, high sales) to the best (high price, low cost, large volume), with the results ranging from a $30,000 loss to a $90,000 profit.

Table 62 Expected Profits Associated with All Possible Combinations of Price, Cost, and Volume

Possible Outcomes				Probability of Occurrence	Expected
Price	Cost	Volume	Profit	(%)	Profit
40	30	1,000	10,000	1.80	180
40	30	2,000	20,000	2.40	480
40	30	3,000	30,000	1.80	540
40	40	1,000	0	3.60	0
40	40	2,000	0	4.80	0
40	40	3,000	0	3.60	0
40	50	1,000	(10,000)	0.60	(60)
40	50	2,000	(20,000)	0.80	(160)
40	50	3,000	(30,000)	0.60	(180)
50	30	1,000	20,000	4.50	900
50	30	2,000	40,000	6.00	2,400
50	30	3,000	60,000	4.50	2,700
50	40	1,000	10,000	9.00	900
50	40	2,000	20,000	12.00	2,400
50	40	3,000	30,000	9.00	2,700
50	50	1,000	0	1.50	0
50	50	2,000	0	2.00	0
50	50	3,000	0	1.50	0
60	30	1,000	30,000	2.70	810
60	30	2,000	60,000	3.60	2,160
60	30	3,000	90,000	2.70	2,430
60	40	1,000	20,000	5.40	1,080
60	40	2,000	40,000	7.20	2,880
60	40	3,000	60,000	5.40	3,240
60	50	1,000	10,000	0.90	90
60	50	2,000	20,000	1.20	240
60	50	3,000	30,000	0.90	270
				100.00	26,000

Each outcome has its own probability of occurrence, as described before. By multiplying the profit to be made in a given situation by the probability of that situation occurring, the expected profit is found for each possible outcome. Then the expected profits are added together for all 27 possible outcomes to arrive at the project's expected profit: $26,000.

While the expected profit is $26,000, there is no way based on the 27 possible combinations of price, cost, and volume that a profit of $26,000 can be achieved.

The value of such an approach is apparent, in that in addition to determining the expected profit for the project, the entire range of possibilities is covered. There is, for example, a 2.0-percent chance that the project will lose money and a 17.0-percent chance that it will just break even. The probability of making some profit therefore is 81.0 percent.

In Table 62A the data from Table 62 are summarized and rearranged in order of increasing profits. The third column in Table 62A gives the cumulative probability distribution, showing the chances of earning *at least* a specified amount. This cumulative distribution is plotted in Figure 40A

Table 62A The Probability of Achieving Various Levels of Profit

Profit or Loss ($)	Probability of Occurrence (%)	Probability of Achieving at Least the Specified Profit (%)
(30,000)	0.6	100.0
(20,000	0.8	99.4
(10,000)	0.6	98.6
0	17.0	98.0
10,000	11.7	81.0
20,000	25.5	69.3
30,000	14.4	43.8
40,000	13.2	29.4
60,000	13.5	16.2
90,000	2.7	2.7

which graphically illustrates the range of possibilities associated with the proposed project.

In situations involving a large number of variables, or in which the range of possible values for the different variables is wide, the total number of combinations can quickly get out of hand. Just five variables each having five possible values will result in a total of 5^5, or 3125, different combinations. While it is obviously impractical to attempt manual calculation of all the possible outcomes in such a case, the problem can easily be handled by a computer.

A computer can quickly calculate and summarize the complete range of possible outcomes, just as was done in the preceding example. Or, it can produce essentially the same results by a process of simulation, in which a large number of possible combinations are drawn at random. The results of the random drawings should, as their number increases, cover the most likely occurrences with approximately the right frequencies, giving results comparable to detailed calculations in far less time.

STATISTICAL MEASUREMENTS

Few exact measurements can be made regarding future expenditures of either time or money. Often only an average value and some indication of minimum and maximum values can be anticipated.

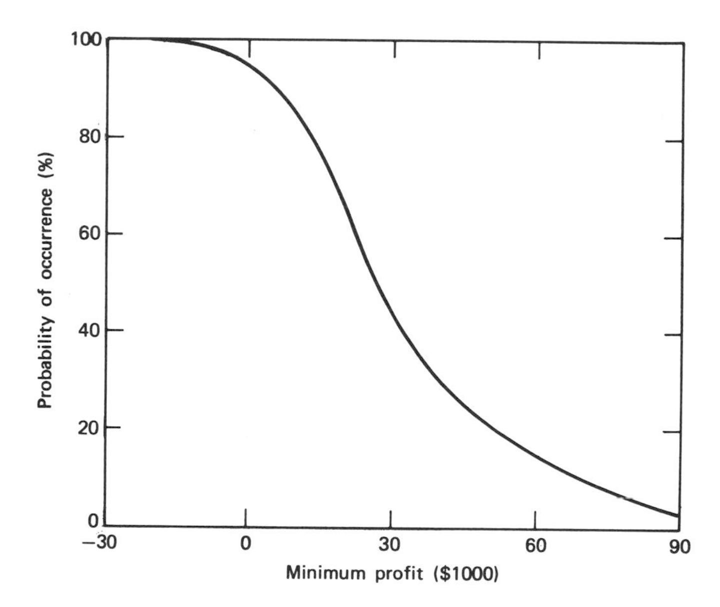

Fig. 40A. *The probability of achieving a specified minimum profit.*

For example, the time required to perform a certain task may vary between 90 and 110 percent of its average time, while another task may vary between 80 and 120 percent. Some way of objectively expressing these variations from the average is therefore desirable so that the variations can be recognized and taken into account when project risks are being evaluated.

The average or mean value of a group of numerical data is found by summing the individual values and then dividing by the number of individual values in the group. In construction scheduling and estimating, the mean or expected value for a specific task is often estimated from just the most optimistic, most pessimistic, and most likely values, as

$$v_e = \frac{a + 4m + b}{6}$$

where v_e is the expected value, a is the minimum value likely to be experienced under normal conditions, b is the maximum value that might be encountered, and m is the most likely value. If the data are normally or symmetrically distributed about their mean, v_e in this formula will be the same as the group mean. In a skewed distribution the expected value will lean in the direction in which there is the greatest uncertainty, or variation from the mean.

This formula can be used safely with most time or cost data. For example, if the most likely total cost of a certain manufacturing process were \$200 per thousand units but might vary between \$175 and \$250 under commonly encountered conditions, the cost to be used for estimating purposes would be:

$$v_e = \frac{175 + 4 \times 200 + 250}{6} = \$204$$

Such an estimate considers the possibility of both better and worse performance than would normally be expected.

VARIANCE. The statistical measure, variance, can be used to describe the reliability of estimates. Variance is defined as the average of the squares of the deviations of a number of observations from their mean value. If the variance is high, the degree of uncertainty associated with the estimate is high; a low variance indicates a high level of accuracy. For rough estimating purposes variance is sometimes considered a function of just the extreme range of values—such as the most optimistic and the most pessimistic values in the preceding example—and can be approximated from the formula:

$$V = \left(\frac{b - a}{6}\right)^2$$

where V is the variance and b and a are the most pessimistic and most optimistic values, respectively. The variance in the manufacturing cost example is found to be

$$V = \left(\frac{250 - 175}{6}\right)^2 = 156$$

While this approach does not give a true measure of statistical variance, it does give a reasonable indication of the relative reliability of various estimates.

STANDARD DEVIATION. Statistically, the standard deviation of a group of data is the square root of the variance; it is defined as the square root of the average of the squares of the differences of the observations from their mean value. The standard deviation is a useful figure to have; if the data are distributed normally, two-thirds of the observations should fall within plus-or-minus one standard deviation of the mean value, and 95 percent will be within plus-or-minus two standard deviations. The smaller the standard deviation, the more reliable and predictable, and less risky, the estimate.

RISK AND UNCERTAINTY

While the terms risk and uncertainty perhaps imply different meanings, their effects are similar. Risks generally refer to situations in which the distribution, or probability of occurrence, of all possible outcomes is either known from past experience or can be calculated with some degree of precision.

Uncertainty, however, covers situations which are of a relatively unique nature and for which the probabilities cannot be calculated.

Risk, then, is simply a measurable uncertainty; uncertainty is a risk that cannot be measured.

Risks may be of either a technical or economic nature. Some risks must be taken; some risks should be taken; some risks are too great to be taken; and other risks may be too good not to take. In any event primary concern should not be with eliminating risks but in selecting the right risks to be taken. The only way to eliminate risks is to do nothing, which is actually the greatest risk of all in business.

Risks in business can never be avoided or eliminated, then, and even minimizing risks is not always desirable. There are several ways, though, that risk and uncertainty can be anticipated and handled so that their possible harmful effects can be minimized.

Risks may be grouped to take advantage of the "law of large numbers," giving the effect of self-insurance. Or, responsibility for assuming risks may be delegated to an insurance company, which in turn consolidates its own risks by grouping the risks of many other individuals or firms.

Some types of risks can actually be controlled or protected against by instigating appropriate protective action. Accident risks, for example, can be controlled by safety measures, and fire and theft can be protected against by installing security controls.

In some cases possible future difficulties can be anticipated on the basis of current information. Increasing the powers of prediction can help to avoid many unpleasant situations, either by changing events now so that the unpleasant events never happen, or by directing the firm's operations along lines that will avoid the less desirable occurrences.

PROFITABILITY VERSUS RISK

The profitability or ROI objective for a new venture should include three major elements: (1) pure interest, (2) compensation for management, and (3) compensation for risk.

The pure interest portion of the return is relatively easy to establish. It represents the return that could be realized by placing the available funds in some alternative, secure, interest-paying investment. This alternative investment might be certificates of deposit, treasury bills, high-grade bonds, or other investment media. In general, the rate of pure interest applicable to invested funds fluctuates between 4 and 8 percent, depending on the condition of the money markets.

Another 1 or 2 percent should be included in the project's target ROI as a reward for management's seeking out, evaluating, and reaching a decision on where the funds could best be placed.

Finally, the risk portion of the project return must be added in. This is strictly a judgment factor and can range from 1 to 40 percent or more depending on the particular project. Typically, a project having average risk should earn from 6 to 10 percent just on the basis of its risk, plus another 4–8 percent for pure interest and a 1–2 percent allowance for investment management. The average project, then, should earn from 11 to 20 percent after taxes, averaging about 15 percent and paying out in about 7 years.

Table 63 summarizes these factors for five categories of projects ranging from very low risk to high risk. Typically, these projects have profitability objectives of 8–40 percent after taxes, resulting in payout periods of between 2.5 and 12 years. Even under the best of conditions, it is difficult to justify projects earning less than 8 percent, or to find projects returning more than 40 percent.

Table 63 Typical Profitability Objectives for Ventures Having Different Levels of Risk

Risk Level	Allowance for Interest (%)	Allowance for Management (%)	Allowance for Risk (%)	Total Return (%)	Typical Return (%)	Typical Payout Time (years)
1 – very low	4 - 8	0 - 1	1 - 3	5 - 12	8	12
2 – below average	4 - 8	1 - 2	2 - 6	7 - 16	11	9
3 – average	4 - 8	1 - 2	6 - 10	11 - 20	15	7
4 – above average	4 - 8	1 - 2	12 - 20	17 - 30	23	4
5 – high	4 - 8	1 - 2	20 - 40	25 - 50	40	2.5

SUMMARY

Statistical techniques are valuable to the engineer in analyzing numerical data for use in predicting or anticipating the outcome of future events. Statistical analysis usually involves the development of frequency distribution data as a basis for probability calculations. Probability theory can be applied in predicting the likelihood of occurrence of uncertain events in numerical terms. The mathematical expectation associated with a variety of alternative situations can be developed by weighing both the probability of an event's occurrence and the potential benefits to be realized should it occur, thereby allowing all possibilities to be recognized and considered and minimizing the need for guesswork on the part of management. Risk and uncertainty are essential parts of business. Risks cannot be eliminated, but their harmful effects can be minimized by thoughtful planning, and business activities can be directed toward areas where risks are most strongly outweighed by potential profits. The profitability objectives for new ventures should be closely related to the risks involved, with high-risk projects requiring a correspondingly high rate of return and rapid payback. Typically, new projects should yield returns of 8–40 percent on invested capital, depending on the degree of risk associated with the venture.

15

Economic and Financial Models

Most engineers are familiar with and experienced in the use of mathematical models in connection with various types of design problems. These mathematical models, whether dealing with the deflection of a column or beam, with the flow of a fluid through a pipe, or with the transmission of electrical energy, are extremely useful—in fact, essential—in presenting an abstract and simplified picture of a realistic physical process.

An engineering model expresses, in a system of mathematical equations, the relationships between several variables and constants. Some variables are explanatory, some are to be explained; some are independent variables, others are dependent variables; some are endogenous variables, others are exogenous variables. Likewise, some of the constants are peculiar to the materials used, while others reflect the external design conditions.

Just as mathematical models are indispensable in engineering studies in giving a simplified picture of what is likely to take place and in focusing attention on the most critical aspects of the situation, economic models are equally useful in feasibility studies.

CHARACTERISTICS OF ECONOMIC MODELS

Economic models define the relationships between several economic variables, expressing these relationships in mathematical terms; the equations usually include both variables and constants, or parameters. But since economic conditions are much less predictable than conditions controlled by the laws of physics, chemistry, and mathematics, economic models often require examination of a wide range of possibilities. Sometimes there are so many different possibilities that the problem is dismissed as being impossible to solve, a most unsatisfactory conclusion. Admittedly, however, in many cases the electronic computer offers the only sensible way of approaching a complex economic problem. The computer performs essentially the same operations in the same sequence as would otherwise be handled manually.

The most valuable use of the computerized economic model is not just in turning out answers but in showing how much difference a change in the assumptions or variables makes in the answer. The actual computer time required to examine different assumptions and recalculate an entire problem is almost insignificant. The computer can in just a few seconds perform calculations that might otherwise require many man-hours of manual effort.

Economic models can be either convenient or flexible, but seldom both at the same time. For convenience and simplicity the model contains mostly constants and only a few variables. For flexibility and realism it probably has a large number of variables and few, if any, constants.

Since profitability is the ultimate goal of any economic enterprise, the determination of profitability is the ultimate objective of most economic models.

FACTORS AFFECTING PROFITS

The major factors that ultimately determine the profits to be made from a business venture, and their relationships to each other, are illustrated in Figure 41.

Profit, the target for any business venture, is directly determined by three factors: cost, price, and volume. Each of these three factors is in turn influenced by other considerations which may be beyond the company's direct control. But whether or not they can be controlled, they must be recognized, understood, and acted upon (or reacted to).

The economic environment determines how large the total market for a specific type of product or service is; competition determines how much of that total market can be obtained at a given price; and the company's efficiency determines how much it costs to produce and market the product or service at the required scale of operations.

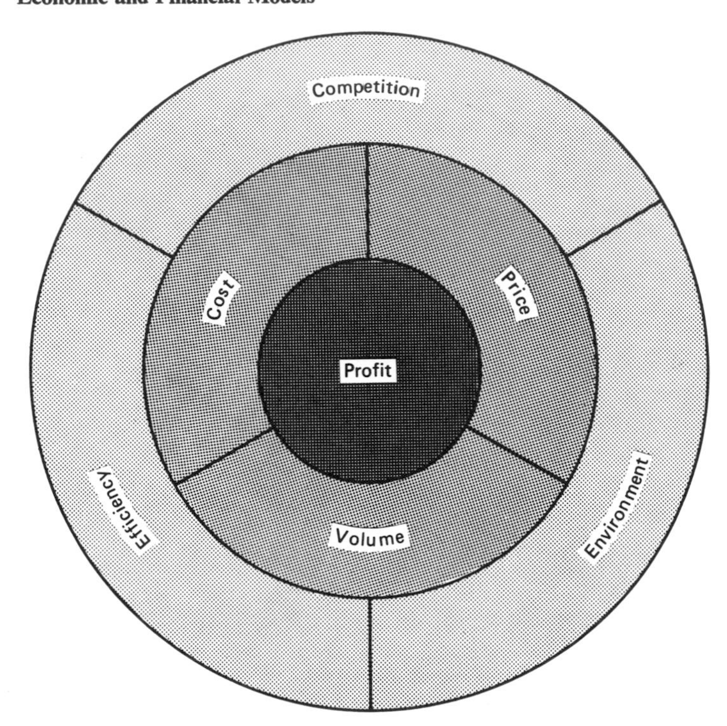

Fig. 41. *Factors affecting profits.*

It is important therefore that operations be planned on the basis of the business climate and outlook, priced on the basis of competitive factors, and carried out as efficiently and economically as possible.

RELATIONSHIPS BETWEEN FACTORS

Price, cost, and volume are the fundamental determinants of a venture's profits; and profits, related to investment, define the venture's profitability.

Several important factors enter into the determination of price, cost, and volume. Some of these factors are shown in Figure 42, where interrelationships between balance sheet and operating statement accounts are illustrated.

The balance sheet items—primarily working capital and fixed investment—constitute the total investment applicable to a given project. From the operating statement, labor, materials, and variable overheads make up the bulk of variable or direct production costs. By applying an appropriate factor (representing capital charges) to fixed investment, and to working capital

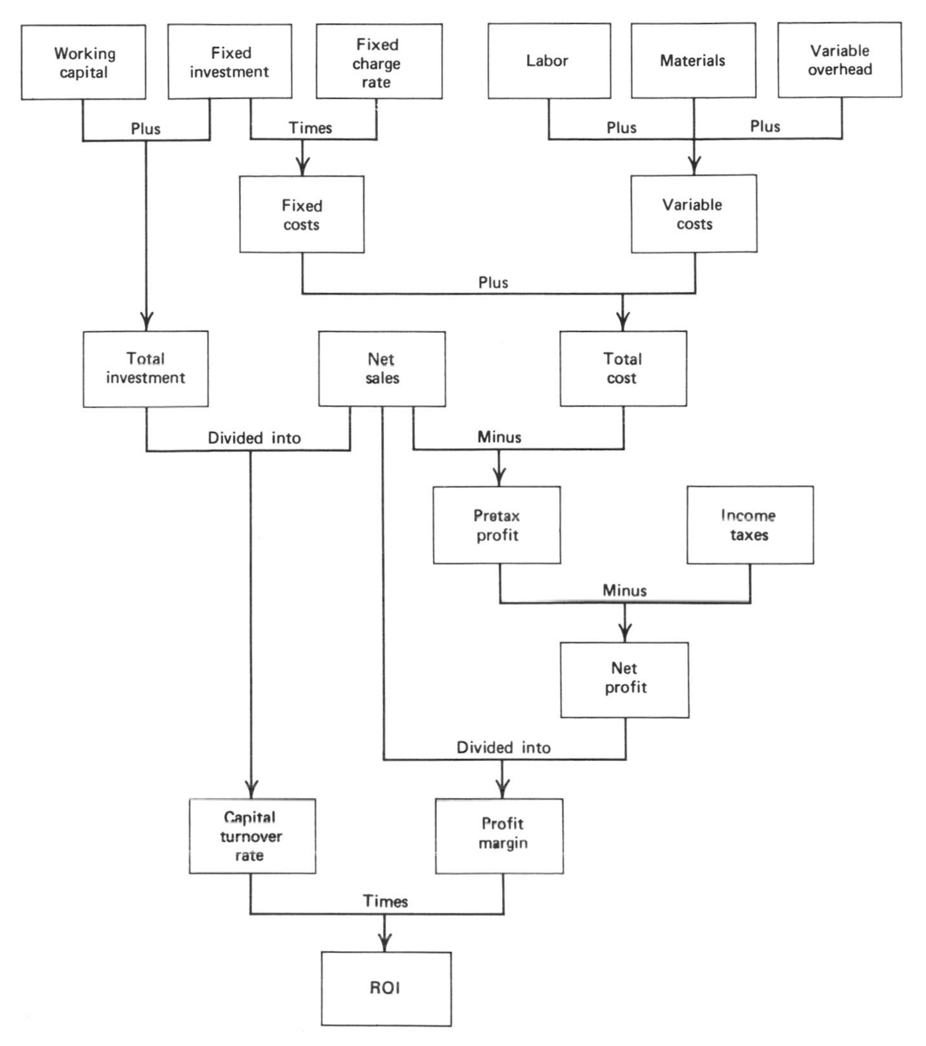

Fig. 42. *Relationships between factors included in the economic model.*

(representing carrying charges), the fixed costs are established and, added to variable costs, give the total cost of doing business under the prescribed conditions.

Net sales less total costs give the pretax profit margin, which can be reduced to net profit by deducting the applicable income taxes.

The resulting investment, sales, and profit figures can then be combined to form several significant financial ratios. Sales divided by investment gives

Table 64 Major Elements of Financial Submodels

Income submodel

    Units sold

x Net price per unit

= Net sales

– Selling costs

= Net income from sales

Manufacturing cost submodel

    Units produced

x Unit cost (materials + labor + variable overhead)

= Direct manufacturing cost

+ Fixed overhead

+ Capital charges

= Total manufacturing cost

Investment submodel

    Depreciable Investment (buildings and equipment)

+ Expensed and amortized investment (engineering, research and development, startup)

+ Land

+ Working capital (receivables and inventory)

= Total capital requirement

Depreciation submodel

    Total depreciable investment

x Depreciation rate

= Annual depreciation charge

Table 64 Major Elements of Financial Submodels (Continued)

Income summary submodel	
	Net income from sales
–	Total manufacturing cost
=	Net operating income
–	Depreciation
–	Expensed investment
=	Net profit before income taxes
–	Income taxes
=	Net profit after taxes
Cash flow submodel	
	Net profit after taxes
+	Depreciation
+	Expensed investment
=	Net cash inflow
–	Net cash outflow (total capital requirement)
=	Net cash flow

the rate of capital turnover, while profit divided by sales gives the profit margin. Finally, the profit margin multiplied by the capital turnover rate gives the return on invested capital:

$$\frac{\text{profit}}{\text{sales}} \times \frac{\text{sales}}{\text{investment}} = \frac{\text{profit}}{\text{investment}}$$

Actually, a profitability model can be broken down into a group of separate and distinct submodels. For example, a complete profitability model might include six separate stages or submodels: a market model, a cost model, an investment model, a depreciation model, a profit model, and a cash flow model. Table 64 summarizes the essential features of the various submodels.

The market, cost, and investment models require considerable work in their development; the other three models simply define the computational procedures to be applied. Finally, the cash flow model is used to develop the appropriate measures of profitability or project desirability.

THE MARKET MODEL

The market model is concerned primarily with the price-volume relationship: how much can be sold at a given price, or what price is required to attain a specific volume. For most products and services, volume increases as the price decreases, and increasing prices are usually accompanied by declining sales. The objective of the market model is to define the probable net income to the company over a period of years, based on various combinations of prices and sales volumes. Price is usually considered the independent variable, and volume the dependent variable; total income is simply the product of price and volume, less direct selling and promotional costs. In analyzing many types of new projects, sales volume can be set at some percent of plant capacity (so many tons per year, rental units available, seating capacity, etc.), and the prices needed to attain that degree of utilization can then be determined based on the market situation. Schedule A (Table 65) shows an example of a sales and income forecast covering a new project over its economic life. In this example the plant will be completed in 2 years, followed by 8 years of production. The product's selling price, initially $1000 per unit, will decrease to $975 per unit when the production capacity of 6000 annual units is reached. Selling costs are estimated to be 10.0 percent of net sales.

THE COST MODEL

The cost-volume relationship is just the opposite of the price-volume relationship. Here, as volume increases, cost decreases; and higher costs are usually associated with lower volumes. The cost model considers direct labor and materials, variable and fixed overheads, and capital charges—all independent variables—to arrive at the dependent variable, the manufacturing cost as a function of volume. If a nonmanufacturing project is involved, only the terminology will be different; instead of manufacturing or production costs, for example, the term operating costs would be more descriptive. Schedule B (Table 66) shows the manufacturing costs for the project whose sales were described in schedule A. Here materials costs and variable overheads are assumed to remain constant on a per-unit basis, while labor costs decrease as production volume increases. The period costs (or time-related charges) stay at $300,000 per year throughout the project's operating life.

Table 65 Schedule A: Sales and Income Forecast

Year	Units Sold	Net Price Per Unit	Net Sales	Selling Costs	Net Income from Sales	
					Per Unit	Annual
0	0					0
1	0					0
2	4,000	1,000	4,000,000	400,000	900.00	3,600,000
3	4,500	1,000	4,500,000	450,000	900.00	4,050,000
4	5,000	1,000	5,000,000	500,000	900.00	4,500,000
5	5,500	1,000	5,500,000	550,000	900.00	4,950,000
6	6,000	975	5,850,000	585,000	877.50	5,265,000
7	6,000	975	5,850,000	585,000	877.50	5,265,000
8	6,000	975	5,850,000	585,000	877.50	5,265,000
9	6,000	975	5,850,000	585,000	877.50	5,265,000

Table 66 Schedule B: Manufacturing Costs

		Unit Costs				Period Costs			Total Manufacturing Cost	
Year	Units Produced	Materials	Labor	Variable Overhead	Total	Fixed Overhead	Capital Charges	Total	Per Unit	Annual
0	0									
1	0									
2	4,000	400.00	60.00	60.00	520.00	50,000	250,000	300,000	595.00	2,380,000
3	4,500	400.00	55.00	60.00	515.00	50,000	250,000	300,000	581.67	2,617,500
4	5,000	400.00	50.00	60.00	510.00	50,000	250,000	300,000	570,00	2,850,000
5	5,500	400.00	45.00	60.00	505.00	50,000	250,000	300,000	559.55	3,077,500
6	6,000	400.00	40.00	60.00	500.00	50,000	250,000	300,000	550.00	3,300,000
7	6,000	400.00	40.00	60.00	500.00	50,000	250,000	300,000	550.00	3,300,000
8	6,000	400.00	40.00	60.00	500.00	50,000	250,000	300,000	550.00	3,300,000
9	6,000	400.00	40.00	60.00	500.00	50,000	250,000	300,000	550.00	3,300,000

THE INVESTMENT MODEL

The magnitude and timing of capital requirements require some careful analytical work, as well as some intuitive judgment. Here a PERT or critical path network proves invaluable in planning the investment schedule. Included in the capital requirements are depreciable investments—consisting primarily of buildings and equipment; expensed or amortized investments such as research and development, engineering, and startup costs; nondepreciable investments such as land and working capital, consisting mainly of the funds needed to carry accounts receivable and inventories. The investment model's output is the total capital required by years over the project's economic life. Schedule C (Table 67) shows the capital requirements by years for the sample project. Year 0 represents the preliminary design and land acquisition phase; year 1 includes most of the actual plant construction, along with additional engineering work; and year 2 sees completion of the plant, followed by startup and initial production. Working capital requirements in the example are estimated at 10.0 percent of total manufacturing cost for inventories and at 10.0 percent of net sales for accounts receivable. The capital requirements for years 3 to 9 represent only the *additional* working capital required each year to keep up with increases in sales and manufacturing costs.

THE DEPRECIATION MODEL

The depreciation model should be set up to handle depreciable assets having different service lives, computing the annual depreciation allowance and the remaining undepreciated balance on an annual basis. The two depreciation methods that allow an accelerated writeoff and are thus favored by most large firms are the sum of the years digits method, and the double declining balance method, switching to a straight-line writeoff at half the asset's estimated life. Appropriate allowances must also be made for investment tax credits and additional first-year depreciation when applicable. Output from the depreciation model consists of total annual depreciation deductions. Schedule D (Table 68) shows the depreciation charges for the example project. Buildings are depreciated over a 20-year period, and equipment over 8 years. Depreciation charges are computed on a double-declining balance basis over the first half of the asset life and then on a straight-line basis for the last half. The rate applied to buildings is 10 percent, and to equipment 25 percent.

THE CASH FLOW MODEL

Except for income tax considerations, the preceding models provide all the data needed for computing net annual cash flows. By assuming an income

Table 67 Schedule C: Capital Requirements

Year	Depreciable Investment		Expensed Investment		Other Capital			Total Capital Requirement
						Working Capital		
	Buildings	Equipment	Engineering R & D	Startup	Land	Receivables	Inventory	
0	0	0	100,000	0	20,000	0	0	120,000
1	500,000	1,200,000	150,000	0		0	0	1,850,000
2	300,000	800,000	50,000	150,000		400,000	238,000	1,938,000
3						450,000	262,000	74,000
4						500,000	285,000	73,000
5						550,000	308,000	73,000
6						585,000	330,000	57,000
7						585,000	330,000	0
8						585,000	330,000	0
9						585,000	330,000	0
Total	800,000	2,000,000	300,000	150,000	20,000	585,000	330,000	4,185,000

Table 68 Schedule D: Depreciation

Year	Building (20 years)		Equipment (8 years)		Total	
	Balance	Amount	Balance	Amount	Balance	Amount
0						
1						
2	800,000	80,000	2,000,000	500,000		580,000
3	720,000	72,000	1,500,000	375,000		447,000
4	648,000	65,000	1,125,000	281,000		346,000
5	583,000	58,000	844,000	211,000		269,000
6	525,000	53,000	633,000	158,000		211,000
7	472,000	47,000	475,000	158,000		205,000
8	425,000	43,000	317,000	158,000		201,000
9	382,000	38,000	159,000	159,000		197,000
10	344,000		0		344,000	

tax rate to be applied to net earnings (about 50 percent, plus whatever tax surcharges may apply, is now typical for large corporations; it is less for small firms), the net annual cash flow, in and out, can be computed over the specified time period. Schedules E and F (Tables 69 and 70) bring together the data from the previous schedules to provide for the computation of net profit after taxes (in schedule E) and cash flow (in schedule F). In schedule E, the net income from sales comes from schedule A, the manufacturing cost from schedule B, depreciation from schedule D, and expensed investment from schedule C. Income taxes are estimated at 50 percent of the net profit before taxes, and all other figures are calculated. It is further assumed that tax losses generated during the early years of the project can be applied to other phases of the company's operations.

In schedule F, cash inflow is calculated by adding depreciation and expensed investment charges (from schedule E) to the net profit after taxes (also from schedule E). Cash outflows are taken directly from schedule C. This results in a net cash outflow for years 0 through 2, followed by 8 years of cash in-

Table 69 Schedule E: Income Statement

Year	Net Income from Sales	Manufacturing Cost	Net Operating Income	Depreciation	Expensed Investment	Net Profit before Taxes	Income Taxes	Net Profit after Taxes
0	0	0	0	0	100,000	(100,000)	(50,000)	(50,000)
1	0	0	0	0	150,000	(150,000)	(75,000)	(75,000)
2	3,600,000	2,380,000	1,220,000	580,000	200,000	440,000	220,000	220,000
3	4,050,000	2,617,500	1,432,500	447,000	0	985,500	492,750	492,750
4	4,500,000	2,850,000	1,650,000	346,000	0	1,304,000	652,000	652,000
5	4,950,000	3,077,500	1,872,500	269,000	0	1,603,500	801,750	801,750
6	5,265,000	3,300,000	1,965,000	211,000	0	1,754,000	877,000	877,000
7	5,265,000	3,300,000	1,965,000	205,000	0	1,760,000	880,000	880,000
8	5,265,000	3,300,000	1,965,000	201,000	0	1,764,000	882,000	882,000
9	5,265,000	3,300,000	1,965,000	197,000	0	1,768,000	884,000	884,000
10	0	0	0	0	0	0	0	0

Table 70 Schedule F: Cash Flow

Year	Net Profit after Taxes	Depreciation and Expensed Investment	Net Cash Inflow	Net Cash Outflow	Net Cash Flow	Cumulative Cash Flow
0	(50,000)	100,000	50,000	120,000	(70,000)	(70,000)
1	(75,000)	150,000	75,000	1,850,000	(1,775,000)	(1,845,000)
2	220,000	780,000	1,000,000	1,938,000	(938,000)	(2,783,000)
3	492,750	447,000	939,750	74,000	865,750	(1,917,250)
4	652,000	346,000	998,000	73,000	925,000	(992,250)
5	801,750	269,000	1,070,750	73,000	997,750	5,500
6	877,000	211,000	1,088,000	57,000	1,031,000	1,036,500
7	880,000	205,000	1,085,000	0	1,085,000	2,121,500
8	882,000	201,000	1,083,000	0	1,083,000	3,204,500
9	884,000	197,000	1,081,000	0	1,081,000	4,285,500
100	0	0	1,279,000	0	1,279,000	5,564,500

flows. In year 10, the inflow shown represents the return of land, working capital, and the undepreciated balance of the building account, upon completion of the project.

Expensed investment is sometimes handled in a different way than shown here. If it is simply treated as an ordinary expense item, it is not included under capital requirements and is not added back to net profits to determine cash flow. The end result is the same in either case; the approach used depends upon the company's accounting preferences.

THE ROI MODEL

From the net annual cash flow and the investment schedules, profitability of the proposed project can be computed by any or all of several methods.

1. The desired rate of return can be set, and the present value of future cash flows can be computed by discounting at the specified rate. This approach can determine if the project is worth pursuing; if the net DCF has a positive value, then the project is earning *at least* the specified rate. In the example, if the cash flows shown in schedule F are discounted at a 10-percent rate, the project will be found to have a present worth of slightly more than $2.0 million. This indicates that, if the company's cost of capital is 10 percent or less, the project is an attractive one.

2. The present value of future cash flows can be set, and the corresponding rate of return (or discount rate) can be computed. If the present value of future cash flows is set at zero, the computed return will be the project's internal rate of return, interest rate of return, profitability index, or DCF rate of return. By employing the DCF technique, the rate of return on invested capital associated with the cash flows summarized in schedule F is about 25 percent. This return is then considered in view of the risks associated with the project, and the investment decision made accordingly. As pointed out in the preceding chapter (see Table 63), a 23-percent return is a typical profitability objective for a venture involving above-average risk. If this were a high-risk project, the return would probably be considered inadequate. For an average-risk project, a 25-percent return is quite attractive.

3. The payout time (or payback period) can be computed, either with or without interest. If the interest rate is set equal to the project's DCF rate of return, the project will exactly pay out over its economic life. In schedule F of the example, it can be seen from the cumulative cash flow that the project will pay out during the fifth year—by interpolation, in about 4.9 years. The reciprocal of this payout time gives another (though poor) measure of investment desirability—about 20 percent in this case.

APPLICATIONS

The profitability model can be used on a project-, product-, plant-, or company-wide basis. Its usefulness can sometimes be enhanced by assigning probabilities to some of the variables, and it can be used in developing optimum price-cost-volume relationships. While the example described in schedules A through F pertains to a manufactured product, the same general format is found adequate for most other types of projects encountered by the engineer.

While many management people are reluctant to acknowledge that such an analytical approach is worthwhile in business economics—perhaps questioning the accuracy implied by any mathematically expressed economic relationships—the real question is not whether this method is infallible but is simply whether there is any other approach as good in formulating assumptions and in evaluating the sensitivity of profits to changes in the variables.

Perhaps the main virtue of this approach, though, is that it requires a great deal of objective thought and study for its preparation. By making the engineer carefully think the project through and state all his assumptions, it helps to minimize the danger of having an individual's preconceived notions lead, after much analysis, only to foregone conclusions.

SUMMARY

The economic or financial model brings together, in summary form, all of the many factors relating to the economic success of a business venture. Profit, a function of price, cost, and volume, is the primary objective of business, so the model necessarily encompasses all the elements that affect price, cost, and volume. The price-volume relationship characterizes the market for a given product and determines the number of units that can be sold at various price levels; the cost-volume relationship identifies economies of scale in production and establishes the costs for whatever level of production is decided upon; and the price-cost relationship defines the net profit margin per unit produced and sold. A comprehensive financial model for a company or for a new venture incorporates all these relationships in separate submodels relating to income, costs, capital requirements, depreciation, profits, and cash flows, which in turn are used to develop appropriate measures of profitability. By relating all the significant factors in a series of mathematical expressions, the impact of changes in any given factor on other factors and on the overall project's economic outcome can be determined. The model permits individual, controllable factors to be varied so that optimum results can be obtained from the project as a whole.

APPENDIX A

Interest Tables

The following interest tables cover 0 to 25 percent in 1.0 percent increments for time periods from 1 to 40 years, for the following factors:

- Single payment compound amount factor.
- Single payment present worth factor.
- Sinking fund factor for uniform annual series.
- Capital recovery factor for uniform annual series.
- Compound amount factor for uniform annual series.
- Present worth factor for uniform annual series.

The following notation has been used: I represents the interest rate per period; N represents the number of interest periods; P represents a present sum of money; S represents a sum of money N interest periods from the present time that, when discounted at interest rate I, is equivalent to a present amount P; R represents the uniform end-of-period payment for N time periods that, discounted at interest rate I, makes the series equivalent to a present amount P.

These tables are reproduced by permission from *Interest Tables: 0 to 25 Percent*, published by the Competitive Service Committee of the Edison Electric Institute (EEI Publ. No. 67-21).

0.00 PERCENT COMPOUND INTEREST FACTORS

N PERIODS	SINGLE PAYMENT: COMPOUND AMOUNT FACTOR GIVEN P TO FIND S	SINGLE PAYMENT: PRESENT WORTH FACTOR GIVEN S TO FIND P	UNIFORM ANNUAL SERIES: SINKING FUND FACTOR GIVEN S TO FIND R	UNIFORM ANNUAL SERIES: CAPITAL RECOVERY FACTOR GIVEN P TO FIND R	UNIFORM ANNUAL SERIES: COMPOUND AMOUNT FACTOR GIVEN R TO FIND S	UNIFORM ANNUAL SERIES: PRESENT WORTH FACTOR GIVEN R TO FIND P	N PERIODS
	(1 + I)**N	1 / (1 + I)**N	I / ((1 + I)**N - 1)	I(1 + I)**N / ((1 + I)**N - 1)	((1 + I)**N - 1) / I	((1 + I)**N - 1) / I(1 + I)**N	
	Equals 1.0 for "N" Periods.	Equals 1.0 for "N" Periods.	Equals 1.0 divided by "N" Periods.	Equals 1.0 divided by "N" Periods.	Equals 1.0 times "N" Periods.	Equals 1.0 times "N" Periods.	

1.00 PERCENT COMPOUND INTEREST FACTORS

N PERIODS	SINGLE PAYMENT: COMPOUND AMOUNT FACTOR GIVEN P TO FIND S	SINGLE PAYMENT: PRESENT WORTH FACTOR GIVEN S TO FIND P	UNIFORM ANNUAL SERIES: SINKING FUND FACTOR GIVEN S TO FIND R	UNIFORM ANNUAL SERIES: CAPITAL RECOVERY FACTOR GIVEN P TO FIND R	UNIFORM ANNUAL SERIES: COMPOUND AMOUNT FACTOR GIVEN R TO FIND S	UNIFORM ANNUAL SERIES: PRESENT WORTH FACTOR GIVEN R TO FIND P	N PERIODS
	(1 + I)**N	1 / (1 + I)**N	I / ((1 + I)**N - 1)	I(1 + I)**N / ((1 + I)**N - 1)	((1 + I)**N - 1) / I	((1 + I)**N - 1) / (I(1 + I)**N)	
1	1.0100000	.9900990	1.0000000	1.0100000	1.0000000	.9900990	1
2	1.0201000	.9802960	.4975124	.5075124	2.0100000	1.9703951	2
3	1.0303010	.9705901	.3300221	.3400221	3.0301000	2.9409852	3
4	1.0406040	.9609803	.2462811	.2562811	4.0604010	3.9019656	4
5	1.0510101	.9514657	.1960398	.2060398	5.1010050	4.8534312	5
6	1.0615202	.9420452	.1625484	.1725484	6.1520151	5.7954765	6
7	1.0721354	.9327181	.1386283	.1486283	7.2135352	6.7281945	7
8	1.0828567	.9234832	.1206903	.1306903	8.2856706	7.6516778	8
9	1.0936853	.9143398	.1067404	.1167404	9.3685273	8.5660176	9
10	1.1046221	.9052870	.0955821	.1055821	10.4622125	9.4713045	10
11	1.1156683	.8963237	.0864541	.0964541	11.5668347	10.3676282	11
12	1.1268250	.8874492	.0788488	.0888488	12.6825030	11.2550775	12
13	1.1380933	.8786626	.0724148	.0824148	13.8093280	12.1337401	13
14	1.1494742	.8699630	.0669012	.0769012	14.9474213	13.0037030	14
15	1.1609690	.8613495	.0621238	.0721238	16.0968955	13.8650525	15

1.00%

16	1.1725786	.8528213	.0579446	.0679446	17.2578645	14.7178738	16
17	1.1843044	.8443775	.0542581	.0642581	18.4304431	15.5622513	17
18	1.1961475	.8360173	.0509820	.0609820	19.6147476	16.3982686	18
19	1.2081090	.8277399	.0480518	.0580518	20.8108950	17.2260085	19
20	1.2201900	.8195445	.0454153	.0554153	22.0190040	18.0455530	20
21	1.2323919	.8114302	.0430308	.0530308	23.2391940	18.8569831	21
22	1.2447159	.8033962	.0408637	.0508637	24.4715860	19.6603793	22
23	1.2571630	.7954418	.0388858	.0488858	25.7163018	20.4558211	23
24	1.2697346	.7875661	.0370735	.0470735	26.9734649	21.2433873	24
25	1.2824320	.7797684	.0354068	.0454068	28.2431995	22.0231557	25
26	1.2952563	.7720480	.0338689	.0438689	29.5256315	22.7952037	26
27	1.3082089	.7644039	.0324455	.0424455	30.8208878	23.5596076	27
28	1.3212910	.7568356	.0311244	.0411244	32.1290967	24.3164432	28
29	1.3345039	.7493421	.0298950	.0398950	33.4503877	25.0657853	29
30	1.3478489	.7419229	.0287481	.0387481	34.7848915	25.8077082	30
31	1.3613274	.7345771	.0276757	.0376757	36.1327404	26.5422854	31
32	1.3749407	.7273041	.0266709	.0366709	37.4940679	27.2695895	32
33	1.3886901	.7201031	.0257274	.0357274	38.8690085	27.9896925	33
34	1.4025770	.7129733	.0248400	.0348400	40.2576986	28.7026659	34
35	1.4166028	.7059142	.0240037	.0340037	41.6602756	29.4085801	35
36	1.4307688	.6989249	.0232143	.0332143	43.0768784	30.1075050	36
37	1.4450765	.6920049	.0224680	.0324680	44.5076471	30.7995099	37
38	1.4595272	.6851534	.0217615	.0317615	45.9527236	31.4846633	38
39	1.4741225	.6783697	.0210916	.0310916	47.4122509	32.1630330	39
40	1.4888637	.6716531	.0204556	.0304556	48.8863734	32.8346861	40

NOTE- **N IS EXPONENT N

2.00 PERCENT COMPOUND INTEREST FACTORS

N PERIODS	SINGLE PAYMENT: COMPOUND AMOUNT FACTOR GIVEN P TO FIND S	SINGLE PAYMENT: PRESENT WORTH FACTOR GIVEN S TO FIND P	UNIFORM ANNUAL SERIES: SINKING FUND FACTOR GIVEN S TO FIND R	UNIFORM ANNUAL SERIES: CAPITAL RECOVERY FACTOR GIVEN P TO FIND R	UNIFORM ANNUAL SERIES: COMPOUND AMOUNT FACTOR GIVEN R TO FIND S	UNIFORM ANNUAL SERIES: PRESENT WORTH FACTOR GIVEN R TO FIND P	N PERIODS
	(1 + I)**N	1 / (1 + I)**N	I / ((1 + I)**N - 1)	I(1 + I)**N / ((1 + I)**N - 1)	((1 + I)**N - 1) / I	((1 + I)**N - 1) / (I(1 + I)**N)	
1	1.0200000	.9803922	1.0000000	1.0200000	1.0000000	.9803922	1
2	1.0404000	.9611688	.4950495	.5150495	2.0200000	1.9415609	2
3	1.0612080	.9423223	.3267547	.3467547	3.0604000	2.8838833	3
4	1.0824322	.9238454	.2426238	.2626238	4.1216080	3.8077287	4
5	1.1040808	.9057308	.1921584	.2121584	5.2040402	4.7134595	5
6	1.1261624	.8879714	.1585258	.1785258	6.3081210	5.6014309	6
7	1.1486857	.8705602	.1345120	.1545120	7.4342834	6.4719911	7
8	1.1716594	.8534904	.1165098	.1365098	8.5829691	7.3254814	8
9	1.1950926	.8367553	.1025154	.1225154	9.7546284	8.1622367	9
10	1.2189944	.8203483	.0913265	.1113265	10.9497210	8.9825850	10
11	1.2433743	.8042630	.0821779	.1021779	12.1687154	9.7868480	11
12	1.2682418	.7884932	.0745596	.0945596	13.4120897	10.5753412	12
13	1.2936066	.7730325	.0681184	.0881184	14.6803315	11.3483737	13
14	1.3194788	.7578750	.0626020	.0826020	15.9739382	12.1062488	14
15	1.3458683	.7430147	.0578255	.0778255	17.2934169	12.8492635	15

2.00%

16	1.3727857	.7284458	.0536501	.0736501	18.6392853	13.5777093	16
17	1.4002414	.7141626	.0499698	.0699698	20.0120710	14.2918719	17
18	1.4282462	.7001594	.0467021	.0667021	21.4123124	14.9920313	18
19	1.4568112	.6864308	.0437818	.0637818	22.8405586	15.6784620	19
20	1.4859474	.6729713	.0411567	.0611567	24.2973698	16.3514333	20
21	1.5156663	.6597758	.0387848	.0587848	25.7833172	17.0112092	21
22	1.5459797	.6468390	.0366314	.0566314	27.2989835	17.6580482	22
23	1.5768993	.6341559	.0346681	.0546681	28.8449632	18.2922041	23
24	1.6084372	.6217215	.0328711	.0528711	30.4218625	18.9139256	24
25	1.6406060	.6095309	.0312204	.0512204	32.0302997	19.5234565	25
26	1.6734181	.5975793	.0296992	.0496992	33.6709057	20.1210358	26
27	1.7068865	.5858620	.0282931	.0482931	35.3443238	20.7068978	27
28	1.7410242	.5743746	.0269897	.0469897	37.0512103	21.2812724	28
29	1.7758447	.5631123	.0257784	.0457784	38.7922345	21.8443847	29
30	1.8113616	.5520709	.0246499	.0446499	40.5680792	22.3964556	30
31	1.8475888	.5412460	.0235963	.0435963	42.3794408	22.9377015	31
32	1.8845406	.5306333	.0226106	.0426106	44.2270296	23.4683348	32
33	1.9222314	.5202287	.0216865	.0416865	46.1115702	23.9885636	33
34	1.9606760	.5100282	.0208187	.0408187	48.0338016	24.4985917	34
35	1.9998896	.5000276	.0200022	.0400022	49.9944776	24.9986193	35
36	2.0398873	.4902232	.0192329	.0392329	51.9943672	25.4888425	36
37	2.0806851	.4806109	.0185068	.0385068	54.0342545	25.9694534	37
38	2.1222988	.4711872	.0178206	.0378206	56.1149396	26.4406406	38
39	2.1647448	.4619482	.0171711	.0371711	58.2372384	26.9025888	39
40	2.2080397	.4528904	.0165557	.0365557	60.4019832	27.3554792	40

NOTE- **N IS EXPONENT N

3.00 PERCENT COMPOUND INTEREST FACTORS

N PERIODS	SINGLE PAYMENT: COMPOUND AMOUNT FACTOR GIVEN P TO FIND S	SINGLE PAYMENT: PRESENT WORTH FACTOR GIVEN S TO FIND P	UNIFORM ANNUAL SERIES: SINKING FUND FACTOR GIVEN S TO FIND R	UNIFORM ANNUAL SERIES: CAPITAL RECOVERY FACTOR GIVEN P TO FIND R	UNIFORM ANNUAL SERIES: COMPOUND AMOUNT FACTOR GIVEN R TO FIND S	UNIFORM ANNUAL SERIES: PRESENT WORTH FACTOR GIVEN R TO FIND P	N PERIODS
	(1 + I)**N	1 / (1 + I)**N	I / ((1 + I)**N - 1)	I(1 + I)**N / ((1 + I)**N - 1)	((1 + I)**N - 1) / I	((1 + I)**N - 1) / I(1 + I)**N	
1	1.0300000	.9708738	1.0000000	1.0300000	1.0000000	.9708738	1
2	1.0609000	.9425959	.4926108	.5226108	2.0300000	1.9134697	2
3	1.0927270	.9151417	.3235304	.3535304	3.0909000	2.8286114	3
4	1.1255088	.8884870	.2390270	.2690270	4.1836270	3.7170984	4
5	1.1592741	.8626088	.1883546	.2183546	5.3091358	4.5797072	5
6	1.1940523	.8374843	.1545975	.1845975	6.4684099	5.4171914	6
7	1.2298739	.8130915	.1305064	.1605064	7.6624622	6.2302830	7
8	1.2667701	.7894092	.1124564	.1424564	8.8923360	7.0196922	8
9	1.3047732	.7664167	.0984339	.1284339	10.1591061	7.7861089	9
10	1.3439164	.7440939	.0872305	.1172305	11.4638793	8.5302028	10
11	1.3842339	.7224213	.0780774	.1080774	12.8077957	9.2526241	11
12	1.4257609	.7013799	.0704621	.1004621	14.1920296	9.9540040	12
13	1.4685337	.6809513	.0640295	.0940295	15.6177904	10.6349553	13
14	1.5125897	.6611178	.0585263	.0885263	17.0863242	11.2960731	14
15	1.5579674	.6418619	.0537666	.0837666	18.5989139	11.9379351	15

3.00%

16	1.6047064	.6231669	.0496108	.0796108	20.1568813	12.5611020	16
17	1.6528476	.6050164	.0459525	.0759525	21.7615877	13.1661185	17
18	1.7024331	.5873946	.0427087	.0727087	23.4144354	13.7535131	18
19	1.7535061	.5702860	.0398139	.0698139	25.1168684	14.3237991	19
20	1.8061112	.5536758	.0372157	.0672157	26.8703745	14.8774749	20
21	1.8602946	.5375493	.0348718	.0648718	28.6764857	15.4150241	21
22	1.9161034	.5218925	.0327474	.0627474	30.5367803	15.9369166	22
23	1.9735865	.5066917	.0308139	.0608139	32.4528837	16.4436084	23
24	2.0327941	.4919337	.0290474	.0590474	34.4264702	16.9355421	24
25	2.0937779	.4776056	.0274279	.0574279	36.4592643	17.4131477	25
26	2.1565913	.4636947	.0259383	.0559383	38.5530423	17.8768424	26
27	2.2212890	.4501891	.0245642	.0545642	40.7096335	18.3270315	27
28	2.2879277	.4370768	.0232932	.0532932	42.9309225	18.7641082	28
29	2.3565655	.4243464	.0221147	.0521147	45.2188502	19.1884546	29
30	2.4272625	.4119868	.0210193	.0510193	47.5754157	19.6004413	30
31	2.5000803	.3999871	.0199989	.0499989	50.0026782	20.0004285	31
32	2.5750828	.3883370	.0190466	.0490466	52.5027585	20.3887655	32
33	2.6523352	.3770262	.0181561	.0481561	55.0778413	20.7657918	33
34	2.7319053	.3660449	.0173220	.0473220	57.7301765	21.1318367	34
35	2.8138625	.3553834	.0165393	.0465393	60.4620818	21.4872201	35
36	2.8982783	.3450324	.0158038	.0458038	63.2759443	21.8322525	36
37	2.9852267	.3349829	.0151116	.0451116	66.1742226	22.1672354	37
38	3.0747835	.3252262	.0144593	.0444593	69.1594493	22.4924616	38
39	3.1670270	.3157535	.0138439	.0438439	72.2342328	22.8082151	39
40	3.2620378	.3065568	.0132624	.0432624	75.4012597	23.1147720	40

NOTE- **N IS EXPONENT N

4.00 PERCENT COMPOUND INTEREST FACTORS

N PERIODS	SINGLE PAYMENT: COMPOUND AMOUNT FACTOR GIVEN P TO FIND S	SINGLE PAYMENT: PRESENT WORTH FACTOR GIVEN S TO FIND P	UNIFORM ANNUAL SERIES: SINKING FUND FACTOR GIVEN S TO FIND R	UNIFORM ANNUAL SERIES: CAPITAL RECOVERY FACTOR GIVEN P TO FIND R	UNIFORM ANNUAL SERIES: COMPOUND AMOUNT FACTOR GIVEN R TO FIND S	UNIFORM ANNUAL SERIES: PRESENT WORTH FACTOR GIVEN R TO FIND P	N PERIODS
	(1 + I)**N	1 / (1 + I)**N	I / ((1 + I)**N - 1)	I(1 + I)**N / ((1 + I)**N - 1)	((1 + I)**N - 1) / I	((1 + I)**N - 1) / I(1 + I)**N	
1	1.0400000	.9615385	1.0000000	1.0400000	1.0000000	.9615385	1
2	1.0816000	.9245562	.4901961	.5301961	2.0400000	1.8860947	2
3	1.1248640	.8889964	.3203485	.3603485	3.1216000	2.7750910	3
4	1.1698586	.8548042	.2354900	.2754900	4.2464640	3.6298952	4
5	1.2166529	.8219271	.1846271	.2246271	5.4163226	4.4518223	5
6	1.2653190	.7903145	.1507619	.1907619	6.6329755	5.2421369	6
7	1.3159318	.7599178	.1266096	.1666096	7.8982945	6.0020547	7
8	1.3685691	.7306902	.1085278	.1485278	9.2142263	6.7327449	8
9	1.4233118	.7025867	.0944930	.1344930	10.5827953	7.4353316	9
10	1.4802443	.6755642	.0832909	.1232909	12.0061071	8.1108958	10
11	1.5394541	.6495809	.0741490	.1141490	13.4863514	8.7604767	11
12	1.6010322	.6245970	.0665522	.1065522	15.0258055	9.3850738	12
13	1.6650735	.6005741	.0601437	.1001437	16.6268377	9.9856478	13
14	1.7316764	.5774751	.0546690	.0946690	18.2919112	10.5631229	14
15	1.8009435	.5552645	.0499411	.0899411	20.0235876	11.1183874	15

4.00%

16	1.8729812	.5339082	.0458200	.0858200	21.8245311	11.6522956	16
17	1.9479005	.5133732	.0421985	.0821985	23.6975124	12.1656689	17
18	2.0258165	.4936281	.0389933	.0789933	25.6454129	12.6592970	18
19	2.1068492	.4746424	.0361386	.0761386	27.6712294	13.1339394	19
20	2.1911231	.4563869	.0335818	.0735818	29.7780786	13.5903263	20
21	2.2787681	.4388336	.0312801	.0712801	31.9692017	14.0291599	21
22	2.3699188	.4219554	.0291988	.0691988	34.2479698	14.4511153	22
23	2.4647155	.4057263	.0273091	.0673091	36.6178886	14.8568417	23
24	2.5633042	.3901215	.0255868	.0655868	39.0826041	15.2469631	24
25	2.6658363	.3751168	.0240120	.0640120	41.6459083	15.6220799	25
26	2.7724698	.3606892	.0225674	.0625674	44.3117446	15.9827692	26
27	2.8833686	.3468166	.0212385	.0612385	47.0842144	16.3295857	27
28	2.9987033	.3334775	.0200130	.0600130	49.9675830	16.6630632	28
29	3.1186515	.3206514	.0188799	.0588799	52.9662863	16.9837146	29
30	3.2433975	.3083187	.0178301	.0578301	56.0849378	17.2920333	30
31	3.3731334	.2964603	.0168554	.0568554	59.3283353	17.5884936	31
32	3.5080587	.2850579	.0159486	.0559486	62.7014687	17.8735515	32
33	3.6483811	.2740942	.0151036	.0551036	66.2095274	18.1476457	33
34	3.7943163	.2635521	.0143148	.0543148	69.8579085	18.4111978	34
35	3.9460890	.2534155	.0135773	.0535773	73.6522249	18.6646132	35
36	4.1039326	.2436687	.0128869	.0528869	77.5983138	18.9082820	36
37	4.2680899	.2342968	.0122396	.0522396	81.7022464	19.1425788	37
38	4.4388135	.2252854	.0116319	.0516319	85.9703363	19.3678642	38
39	4.6163660	.2166206	.0110608	.0510608	90.4091497	19.5844848	39
40	4.8010206	.2082890	.0105235	.0505235	95.0255157	19.7927739	40

NOTE- **N IS EXPONENT N

5.00 PERCENT COMPOUND INTEREST FACTORS

N PERIODS	SINGLE PAYMENT — COMPOUND AMOUNT FACTOR GIVEN P TO FIND S	SINGLE PAYMENT — PRESENT WORTH FACTOR GIVEN S TO FIND P	UNIFORM ANNUAL SERIES — SINKING FUND FACTOR GIVEN S TO FIND R	UNIFORM ANNUAL SERIES — CAPITAL RECOVERY FACTOR GIVEN P TO FIND R	UNIFORM ANNUAL SERIES — COMPOUND AMOUNT FACTOR GIVEN R TO FIND S	UNIFORM ANNUAL SERIES — PRESENT WORTH FACTOR GIVEN R TO FIND P	N PERIODS
	(1 + I)**N	1 / (1 + I)**N	I / ((1 + I)**N - 1)	I(1 + I)**N / ((1 + I)**N - 1)	((1 + I)**N - 1) / I	((1 + I)**N - 1) / (I(1 + I)**N)	
1	1.0500000	.9523810	1.0000000	1.0500000	1.0000000	.9523810	1
2	1.1025000	.9070295	.4878049	.5378049	2.0500000	1.8594104	2
3	1.1576250	.8638376	.3172086	.3672086	3.1525000	2.7232480	3
4	1.2155063	.8227025	.2320118	.2820118	4.3101250	3.5459505	4
5	1.2762816	.7835262	.1809748	.2309748	5.5256313	4.3294767	5
6	1.3400956	.7462154	.1470175	.1970175	6.8019128	5.0756921	6
7	1.4071004	.7106813	.1228198	.1728198	8.1420085	5.7863734	7
8	1.4774554	.6768394	.1047218	.1547218	9.5491089	6.4632128	8
9	1.5513282	.6446089	.0906901	.1406901	11.0265643	7.1078217	9
10	1.6288946	.6139133	.0795046	.1295046	12.5778925	7.7217349	10
11	1.7103394	.5846793	.0703889	.1203889	14.2067872	8.3064142	11
12	1.7958563	.5568374	.0628254	.1128254	15.9171265	8.8632516	12
13	1.8856491	.5303214	.0564558	.1064558	17.7129828	9.3935730	13
14	1.9799316	.5050680	.0510240	.1010240	19.5986320	9.8986409	14
15	2.0789282	.4810171	.0463423	.0963423	21.5785636	10.3796580	15

5.00%

16	2.1828746	.4581115	.0422699	.0922699	23.6574918	10.8377696	16
17	2.2920183	.4362967	.0386991	.0886991	25.8403664	11.2740662	17
18	2.4066192	.4155207	.0355462	.0855462	28.1323847	11.6895869	18
19	2.5269502	.3957340	.0327450	.0827450	30.5390039	12.0853209	19
20	2.6532977	.3768895	.0302426	.0802426	33.0659541	12.4622103	20
21	2.7859626	.3589424	.0279961	.0779961	35.7192518	12.8211527	21
22	2.9252607	.3418499	.0259705	.0759705	38.5052144	13.1630026	22
23	3.0715238	.3255713	.0241368	.0741368	41.4304751	13.4885739	23
24	3.2250999	.3100679	.0224709	.0724709	44.5019989	13.7986418	24
25	3.3863549	.2953028	.0209525	.0709525	47.7270988	14.0939446	25
26	3.5556727	.2812407	.0195643	.0695643	51.1134538	14.3751853	26
27	3.7334563	.2678483	.0182919	.0682919	54.6691264	14.6430336	27
28	3.9201291	.2550936	.0171225	.0671225	58.4025828	14.8981273	28
29	4.1161356	.2429463	.0160455	.0660455	62.3227119	15.1410736	29
30	4.3219424	.2313774	.0150514	.0650514	66.4388475	15.3724510	30
31	4.5380395	.2203595	.0141321	.0641321	70.7607899	15.5928105	31
32	4.7649415	.2098662	.0132804	.0632804	75.2988294	15.8026767	32
33	5.0031885	.1998725	.0124900	.0624900	80.0637708	16.0025492	33
34	5.2533480	.1903548	.0117554	.0617554	85.0669594	16.1929040	34
35	5.5160154	.1812903	.0110717	.0610717	90.3203074	16.3741943	35
36	5.7918161	.1726574	.0104345	.0604345	95.8363227	16.5468517	36
37	6.0814069	.1644356	.0098398	.0598398	101.6281389	16.7112873	37
38	6.3854773	.1566054	.0092842	.0592842	107.7095458	16.8678927	38
39	6.7047512	.1491480	.0087646	.0587646	114.0950231	17.0170407	39
40	7.0399887	.1420457	.0082782	.0582782	120.7997742	17.1590864	40

NOTE- **N IS EXPONENT N

6.00 PERCENT COMPOUND INTEREST FACTORS

N PERIODS	SINGLE PAYMENT COMPOUND AMOUNT FACTOR GIVEN P TO FIND S	SINGLE PAYMENT PRESENT WORTH FACTOR GIVEN S TO FIND P	UNIFORM ANNUAL SERIES SINKING FUND FACTOR GIVEN S TO FIND R	UNIFORM ANNUAL SERIES CAPITAL RECOVERY FACTOR GIVEN P TO FIND R	UNIFORM ANNUAL SERIES COMPOUND AMOUNT FACTOR GIVEN R TO FIND S	UNIFORM ANNUAL SERIES PRESENT WORTH FACTOR GIVEN R TO FIND P	N PERIODS
	(1 + I)**N	1 / (1 + I)**N	I / ((1 + I)**N - 1)	I(1 + I)**N / ((1 + I)**N - 1)	((1 + I)**N - 1) / I	((1 + I)**N - 1) / (I(1 + I)**N)	
1	1.0600000	.9433962	1.0000000	1.0600000	1.0000000	.9433962	1
2	1.1236000	.8899964	.4854369	.5454369	2.0600000	1.8333927	2
3	1.1910160	.8396193	.3141098	.3741098	3.1836000	2.6730119	3
4	1.2624770	.7920937	.2285915	.2885915	4.3746160	3.4651056	4
5	1.3382256	.7472582	.1773964	.2373964	5.6370930	4.2123638	5
6	1.4185191	.7049605	.1433626	.2033626	6.9753185	4.9173243	6
7	1.5036303	.6650571	.1191350	.1791350	8.3938376	5.5823814	7
8	1.5938481	.6274124	.1010359	.1610359	9.8974679	6.2097938	8
9	1.6894790	.5918985	.0870222	.1470222	11.4913160	6.8016923	9
10	1.7908477	.5583948	.0758680	.1358680	13.1807949	7.3600871	10
11	1.8982986	.5267875	.0667929	.1267929	14.9716426	7.8868746	11
12	2.0121965	.4969694	.0592770	.1192770	16.8699412	8.3838439	12
13	2.1329283	.4688390	.0529601	.1129601	18.8821377	8.8526830	13
14	2.2609040	.4423010	.0475849	.1075849	21.0150659	9.2949839	14
15	2.3965582	.4172651	.0429628	.1029628	23.2759699	9.7122490	15

6.00%

16	2.5403517	.3936463	.0389521	.0989521	25.6725281	10.1058953	16
17	2.6927728	.3713644	.0354448	.0954448	28.2128798	10.4772597	17
18	2.8543392	.3503438	.0323565	.0923565	30.9056525	10.8276035	18
19	3.0255995	.3305130	.0296209	.0896209	33.7599917	11.1581165	19
20	3.2071355	.3118047	.0271846	.0871846	36.7855912	11.4699212	20
21	3.3995636	.2941554	.0250045	.0850045	39.9927267	11.7640766	21
22	3.6035374	.2775051	.0230456	.0830456	43.3922903	12.0415817	22
23	3.8197497	.2617973	.0212785	.0812785	46.9958277	12.3033790	23
24	4.0489346	.2469785	.0196790	.0796790	50.8155774	12.5503575	24
25	4.2918707	.2329986	.0182267	.0782267	54.8645120	12.7833562	25
26	4.5493830	.2198100	.0169043	.0769043	59.1563827	13.0031662	26
27	4.8223459	.2073680	.0156972	.0756972	63.7057657	13.2105341	27
28	5.1116867	.1956301	.0145926	.0745926	68.5281116	13.4061643	28
29	5.4183879	.1845567	.0135796	.0735796	73.6397983	13.5907210	29
30	5.7434912	.1741101	.0126489	.0726489	79.0581862	13.7648312	30
31	6.0881006	.1642548	.0117922	.0717922	84.8016774	13.9290860	31
32	6.4533867	.1549574	.0110023	.0710023	90.8897780	14.0840434	32
33	6.8405899	.1461862	.0102729	.0702729	97.3431647	14.2302296	33
34	7.2510253	.1379115	.0095984	.0695984	104.1837546	14.3681411	34
35	7.6860868	.1301052	.0089739	.0689739	111.4347799	14.4982464	35
36	8.1472520	.1227408	.0083948	.0683948	119.1208667	14.6209871	36
37	8.6360871	.1157932	.0078574	.0678574	127.2681187	14.7367803	37
38	9.1542523	.1092389	.0073581	.0673581	135.9042058	14.8460192	38
39	9.7035075	.1030555	.0068938	.0668938	145.0584581	14.9490747	39
40	10.2857179	.0972222	.0064615	.0664615	154.7619656	15.0462969	40

NOTE- **N IS EXPONENT N

7.00 PERCENT COMPOUND INTEREST FACTORS

N PERIODS	SINGLE PAYMENT: COMPOUND AMOUNT FACTOR GIVEN P TO FIND S	SINGLE PAYMENT: PRESENT WORTH FACTOR GIVEN S TO FIND P	UNIFORM ANNUAL SERIES: SINKING FUND FACTOR GIVEN S TO FIND R	UNIFORM ANNUAL SERIES: CAPITAL RECOVERY FACTOR GIVEN P TO FIND R	UNIFORM ANNUAL SERIES: COMPOUND AMOUNT FACTOR GIVEN R TO FIND S	UNIFORM ANNUAL SERIES: PRESENT WORTH FACTOR GIVEN R TO FIND P	N PERIODS
	(1 + I)**N	1 / (1 + I)**N	I / ((1 + I)**N - 1)	I(1 + I)**N / ((1 + I)**N - 1)	((1 + I)**N - 1) / I	((1 + I)**N - 1) / (I(1 + I)**N)	
1	1.0700000	.9345794	1.0000000	1.0700000	1.0000000	.9345794	1
2	1.1449000	.8734387	.4830918	.5530918	2.0700000	1.8080182	2
3	1.2250430	.8162979	.3110517	.3810517	3.2149000	2.6243160	3
4	1.3107960	.7628952	.2252281	.2952281	4.4399430	3.3872113	4
5	1.4025517	.7129862	.1738907	.2438907	5.7507390	4.1001974	5
6	1.5007304	.6663422	.1397958	.2097958	7.1532907	4.7665397	6
7	1.6057815	.6227497	.1155532	.1855532	8.6540211	5.3892894	7
8	1.7181862	.5820091	.0974678	.1674678	10.2598026	5.9712985	8
9	1.8384592	.5439337	.0834865	.1534865	11.9779887	6.5152322	9
10	1.9671514	.5083493	.0723775	.1423775	13.8164480	7.0235815	10
11	2.1048520	.4750928	.0633569	.1333569	15.7835993	7.4986743	11
12	2.2521916	.4440120	.0559020	.1259020	17.8884513	7.9426863	12
13	2.4098450	.4149644	.0496508	.1196508	20.1406429	8.3576507	13
14	2.5785342	.3878172	.0443449	.1143449	22.5504879	8.7454680	14
15	2.7590315	.3624460	.0397946	.1097946	25.1290220	9.1079140	15

7.00%

16	2.9521637	.3387346	.0358576	.1058576	27.8880536	9.4466486	16
17	3.1588152	.3165744	.0324252	.1024252	30.8402173	9.7632230	17
18	3.3799323	.2958639	.0294126	.0994126	33.9990325	10.0590869	18
19	3.6165275	.2765083	.0267530	.0967530	37.3789648	10.3355952	19
20	3.8696845	.2584190	.0243929	.0943929	40.9954923	10.5940142	20
21	4.1405624	.2415131	.0222890	.0922890	44.8651768	10.8355273	21
22	4.4304017	.2257132	.0204058	.0904058	49.0057392	11.0612405	22
23	4.7405299	.2109469	.0187139	.0887139	53.4361409	11.2721874	23
24	5.0723670	.1971466	.0171890	.0871890	58.1766708	11.4693340	24
25	5.4274326	.1842492	.0158105	.0858105	63.2490377	11.6535832	25
26	5.8073529	.1721955	.0145610	.0845610	68.6764704	11.8257787	26
27	6.2138676	.1609304	.0134257	.0834257	74.4838233	11.9867090	27
28	6.6488384	.1504022	.0123919	.0823919	80.6976909	12.1371113	28
29	7.1142570	.1405628	.0114487	.0814487	87.3465293	12.2776741	29
30	7.6122550	.1313671	.0105864	.0805864	94.4607863	12.4090412	30
31	8.1451129	.1227730	.0097969	.0797969	102.0730414	12.5318142	31
32	8.7152708	.1147411	.0090729	.0790729	110.2181543	12.6465553	32
33	9.3253398	.1072347	.0084081	.0784081	118.9334251	12.7537900	33
34	9.9781135	.1002193	.0077967	.0777967	128.2587648	12.8540094	34
35	10.6765815	.0936629	.0072340	.0772340	138.2368784	12.9476723	35
36	11.4239422	.0875355	.0067153	.0767153	148.9134598	13.0352078	36
37	12.2236181	.0818088	.0062368	.0762368	160.3374020	13.1170166	37
38	13.0792714	.0764569	.0057951	.0757951	172.5610202	13.1934735	38
39	13.9948204	.0714550	.0053868	.0753868	185.6402916	13.2649285	39
40	14.9744578	.0667804	.0050091	.0750091	199.6351120	13.3317088	40

NOTE- **N IS EXPONENT N

8.00 PERCENT COMPOUND INTEREST FACTORS

N PERIODS	SINGLE PAYMENT: COMPOUND AMOUNT FACTOR GIVEN P TO FIND S	SINGLE PAYMENT: PRESENT WORTH FACTOR GIVEN S TO FIND P	UNIFORM ANNUAL SERIES: SINKING FUND FACTOR GIVEN S TO FIND R	UNIFORM ANNUAL SERIES: CAPITAL RECOVERY FACTOR GIVEN P TO FIND R	UNIFORM ANNUAL SERIES: COMPOUND AMOUNT FACTOR GIVEN R TO FIND S	UNIFORM ANNUAL SERIES: PRESENT WORTH FACTOR GIVEN R TO FIND P	N PERIODS
	(1 + I)**N	1 / (1 + I)**N	I / ((1 + I)**N - 1)	I(1 + I)**N / ((1 + I)**N - 1)	((1 + I)**N - 1) / I	((1 + I)**N - 1) / I(1 + I)**N	
1	1.0800000	.9259259	1.0000000	1.0800000	1.0000000	.9259259	1
2	1.1664000	.8573388	.4807692	.5607692	2.0800000	1.7832647	2
3	1.2597120	.7938322	.3080335	.3880335	3.2464000	2.5770970	3
4	1.3604890	.7350299	.2219208	.3019208	4.5061120	3.3121268	4
5	1.4693281	.6805832	.1704565	.2504565	5.8666010	3.9927100	5
6	1.5868743	.6301696	.1363154	.2163154	7.3359290	4.6228797	6
7	1.7138243	.5834904	.1120724	.1920724	8.9228034	5.2063701	7
8	1.8509302	.5402689	.0940148	.1740148	10.6366276	5.7466389	8
9	1.9990046	.5002490	.0800797	.1600797	12.4875578	6.2468879	9
10	2.1589250	.4631935	.0690295	.1490295	14.4865625	6.7100814	10
11	2.3316390	.4288829	.0600763	.1400763	16.6454875	7.1389643	11
12	2.5181701	.3971138	.0526950	.1326950	18.9771265	7.5360780	12
13	2.7196237	.3676979	.0465218	.1265218	21.4952966	7.9037759	13
14	2.9371936	.3404610	.0412969	.1212969	24.2149203	8.2442370	14
15	3.1721691	.3152417	.0368295	.1168295	27.1521139	8.5594787	15

8.00%

16	3.4259426	.2918905	.0329769	.1129769	30.3242830	8.8513692	16
17	3.7000181	.2702690	.0296294	.1096294	33.7502257	9.1216381	17
18	3.9960195	.2502490	.0267021	.1067021	37.4502437	9.3718871	18
19	4.3157011	.2317121	.0241276	.1041276	41.4462632	9.6035992	19
20	4.6609571	.2145482	.0218522	.1018522	45.7619643	9.8181474	20
21	5.0338337	.1986557	.0198323	.0998323	50.4229214	10.0168032	21
22	5.4365404	.1839405	.0180321	.0980321	55.4567552	10.2007437	22
23	5.8714636	.1703153	.0164222	.0964222	60.8932956	10.3710589	23
24	6.3411807	.1576993	.0149780	.0949780	66.7647592	10.5287583	24
25	6.8484752	.1460179	.0136788	.0936788	73.1059400	10.6747762	25
26	7.3963532	.1352018	.0125071	.0925071	79.9544151	10.8099780	26
27	7.9880615	.1251868	.0114481	.0914481	87.3507684	10.9351648	27
28	8.6271064	.1159137	.0104889	.0904889	95.3388298	11.0510785	28
29	9.3172749	.1073275	.0096185	.0896185	103.9659362	11.1584060	29
30	10.0626569	.0993773	.0088274	.0888274	113.2832111	11.2577833	30
31	10.8676694	.0920160	.0081073	.0881073	123.3458680	11.3497994	31
32	11.7370830	.0852000	.0074508	.0874508	134.2135374	11.4349994	32
33	12.6760496	.0788889	.0068516	.0868516	145.9506204	11.5138884	33
34	13.6901336	.0730453	.0063041	.0863041	158.6266701	11.5869337	34
35	14.7853443	.0676345	.0058033	.0858033	172.3168037	11.6545682	35
36	15.9681718	.0626246	.0053447	.0853447	187.1021480	11.7171928	36
37	17.2456256	.0579857	.0049244	.0849244	203.0703198	11.7751785	37
38	18.6252756	.0536905	.0045389	.0845389	220.3159454	11.8288690	38
39	20.1152977	.0497134	.0041851	.0841851	238.9412210	11.8785824	39
40	21.7245215	.0460309	.0038602	.0838602	259.0565187	11.9246133	40

NOTE- **N IS EXPONENT N

9.00 PERCENT COMPOUND INTEREST FACTORS

N PERIODS	SINGLE PAYMENT: COMPOUND AMOUNT FACTOR GIVEN P TO FIND S	SINGLE PAYMENT: PRESENT WORTH FACTOR GIVEN S TO FIND P	UNIFORM ANNUAL SERIES: SINKING FUND FACTOR GIVEN S TO FIND R	UNIFORM ANNUAL SERIES: CAPITAL RECOVERY FACTOR GIVEN P TO FIND R	UNIFORM ANNUAL SERIES: COMPOUND AMOUNT FACTOR GIVEN R TO FIND S	UNIFORM ANNUAL SERIES: PRESENT WORTH FACTOR GIVEN R TO FIND P	N PERIODS
	(1 + I)**N	1 / (1 + I)**N	I / ((1 + I)**N - 1)	I(1 + I)**N / ((1 + I)**N - 1)	((1 + I)**N - 1) / I	((1 + I)**N - 1) / I(1 + I)**N	
1	1.0900000	.9174312	1.0000000	1.0900000	1.0000000	.9174312	1
2	1.1881000	.8416800	.4784689	.5684689	2.0900000	1.7591112	2
3	1.2950290	.7721835	.3050548	.3950548	3.2781000	2.5312947	3
4	1.4115816	.7084252	.2186687	.3086687	4.5731290	3.2397199	4
5	1.5386240	.6499314	.1670925	.2570925	5.9847106	3.8896513	5
6	1.6771001	.5962673	.1329198	.2229198	7.5233346	4.4859186	6
7	1.8280391	.5470342	.1086905	.1986905	9.2004347	5.0329528	7
8	1.9925626	.5018663	.0906744	.1806744	11.0284738	5.5348191	8
9	2.1718933	.4604278	.0767988	.1667988	13.0210364	5.9952469	9
10	2.3673637	.4224108	.0658201	.1558201	15.1929297	6.4176577	10
11	2.5804264	.3875329	.0569467	.1469467	17.5602934	6.8051906	11
12	2.8126648	.3555347	.0496507	.1396507	20.1407198	7.1607253	12
13	3.0658046	.3261786	.0435666	.1335666	22.9533846	7.4869039	13
14	3.3417270	.2992465	.0384332	.1284332	26.0191892	7.7861504	14
15	3.6424825	.2745380	.0340589	.1240589	29.3609162	8.0606884	15

9.00%

16	3.9703059	.2518698	.0302999	.1202999	33.0033987	8.3125582	16
17	4.3276334	.2310732	.0270462	.1170462	36.9737046	8.5436314	17
18	4.7171204	.2119937	.0242123	.1142123	41.3013380	8.7556251	18
19	5.1416613	.1944897	.0217304	.1117304	46.0184584	8.9501148	19
20	5.6044108	.1784309	.0195465	.1095465	51.1601196	9.1285457	20
21	5.1088077	.1636981	.0176166	.1076166	56.7645304	9.2922437	21
22	5.6586004	.1501817	.0159050	.1059050	62.8733381	9.4424254	22
23	7.2578745	.1377814	.0143819	.1043819	69.5319386	9.5802068	23
24	7.9110832	.1264049	.0130226	.1030226	76.7898131	9.7066118	24
25	3.6230807	.1159678	.0118063	.1018063	84.7008962	9.8225796	25
26	9.3991579	.1063925	.0107154	.1007154	93.3239769	9.9289721	26
27	10.2450821	.0976078	.0097349	.0997349	102.7231348	10.0265799	27
28	11.1671395	.0895484	.0088520	.0988520	112.9682169	10.1161284	28
29	12.1721821	.0821545	.0080557	.0980557	124.1353565	10.1982829	29
30	13.2676785	.0753711	.0073364	.0973364	136.3075385	10.2736540	30
31	14.4617695	.0691478	.0066856	.0966856	149.5752170	10.3428019	31
32	15.7633288	.0634384	.0060962	.0960962	164.0369865	10.4062403	32
33	17.1820284	.0582003	.0055617	.0955617	179.8003153	10.4644406	33
34	13.7284109	.0533948	.0050766	.0950766	196.9823437	10.5178354	34
35	20.4139679	.0489861	.0046358	.0946358	215.7107547	10.5668215	35
36	22.2512250	.0449413	.0042350	.0942350	236.1247226	10.6117628	36
37	24.2538353	.0412306	.0038703	.0938703	258.3759476	10.6529934	37
38	25.4366805	.0378262	.0035382	.0935382	282.6297829	10.6908196	38
39	23.8159817	.0347030	.0032356	.0932356	309.0664633	10.7255226	39
40	31.4094201	.0318376	.0029596	.0929596	337.8824450	10.7573602	40

NOTE- **N IS EXPONENT N

10.00 PERCENT COMPOUND INTEREST FACTORS

N PERIODS	SINGLE PAYMENT: COMPOUND AMOUNT FACTOR GIVEN P TO FIND S	SINGLE PAYMENT: PRESENT WORTH FACTOR GIVEN S TO FIND P	UNIFORM ANNUAL SERIES: SINKING FUND FACTOR GIVEN S TO FIND R	UNIFORM ANNUAL SERIES: CAPITAL RECOVERY FACTOR GIVEN P TO FIND R	UNIFORM ANNUAL SERIES: COMPOUND AMOUNT FACTOR GIVEN R TO FIND S	UNIFORM ANNUAL SERIES: PRESENT WORTH FACTOR GIVEN R TO FIND P	N PERIODS
	(1 + I)**N	1 / (1 + I)**N	I / ((1 + I)**N - 1)	I(1 + I)**N / ((1 + I)**N - 1)	((1 + I)**N - 1) / I	((1 + I)**N - 1) / (I(1 + I)**N)	
1	1.1000000	.9090909	1.0000000	1.1000000	1.0000000	.9090909	1
2	1.2100000	.8264463	.4761905	.5761905	2.1000000	1.7355372	2
3	1.3310000	.7513148	.3021148	.4021148	3.3100000	2.4868520	3
4	1.4641000	.6830135	.2154708	.3154708	4.6410000	3.1698654	4
5	1.6105100	.6209213	.1637975	.2637975	6.1051000	3.7907868	5
6	1.7715610	.5644739	.1296074	.2296074	7.7156100	4.3552607	6
7	1.9487171	.5131581	.1054055	.2054055	9.4871710	4.8684188	7
8	2.1435888	.4665074	.0874440	.1874440	11.4358881	5.3349262	8
9	2.3579477	.4240976	.0736405	.1736405	13.5794769	5.7590238	9
10	2.5937425	.3855433	.0627454	.1627454	15.9374246	6.1445671	10
11	2.8531167	.3504939	.0539631	.1539631	18.5311671	6.4950610	11
12	3.1384284	.3186308	.0467633	.1467633	21.3842838	6.8136918	12
13	3.4522712	.2896644	.0407785	.1407785	24.5227121	7.1033562	13
14	3.7974983	.2633313	.0357462	.1357462	27.9749834	7.3666875	14
15	4.1772482	.2393920	.0314738	.1314738	31.7724817	7.6060795	15

10.00%

16	4.5949730	.2176291	.0278166	.1278166	35.9497299	7.8237086	16
17	5.0544703	.1978447	.0246641	.1246641	40.5447028	8.0215533	17
18	5.5599173	.1798588	.0219302	.1219302	45.5991731	8.2014121	18
19	6.1159090	.1635080	.0195469	.1195469	51.1590904	8.3649201	19
20	6.7274999	.1486436	.0174596	.1174596	57.2749995	8.5135637	20
21	7.4002499	.1351306	.0156244	.1156244	64.0024994	8.6486943	21
22	8.1402749	.1228460	.0140051	.1140051	71.4027494	8.7715403	22
23	8.9543024	.1116782	.0125718	.1125718	79.5430243	8.8832184	23
24	9.8497327	.1015256	.0112998	.1112998	88.4973268	8.9847440	24
25	10.8347059	.0922960	.0101681	.1101681	98.3470594	9.0770400	25
26	11.9181765	.0839055	.0091590	.1091590	109.1817654	9.1609455	26
27	13.1099942	.0762777	.0082576	.1082576	121.0999419	9.2372232	27
28	14.4209936	.0693433	.0074510	.1074510	134.2099361	9.3065665	28
29	15.8630930	.0630394	.0067281	.1067281	148.6309297	9.3696059	29
30	17.4494023	.0573086	.0060792	.1060792	164.4940227	9.4269145	30
31	19.1943425	.0520987	.0054962	.1054962	181.9434250	9.4790132	31
32	21.1137767	.0473624	.0049717	.1049717	201.1377675	9.5263756	32
33	23.2251544	.0430568	.0044994	.1044994	222.2515442	9.5694324	33
34	25.5476699	.0391425	.0040737	.1040737	245.4766986	9.6085749	34
35	28.1024368	.0355841	.0036897	.1036897	271.0243685	9.6441590	35
36	30.9126805	.0323492	.0033431	.1033431	299.1268053	9.6765082	36
37	34.0039486	.0294083	.0030299	.1030299	330.0394859	9.7059165	37
38	37.4043434	.0267349	.0027469	.1027469	364.0434344	9.7326514	38
39	41.1447778	.0243044	.0024910	.1024910	401.4477779	9.7569558	39
40	45.2592556	.0220949	.0022594	.1022594	442.5925557	9.7790507	40

NOTE- **N IS EXPONENT N

11.00 PERCENT COMPOUND INTEREST FACTORS

	SINGLE PAYMENT		UNIFORM ANNUAL SERIES				
N PERIODS	COMPOUND AMOUNT FACTOR GIVEN P TO FIND S	PRESENT WORTH FACTOR GIVEN S TO FIND P	SINKING FUND FACTOR GIVEN S TO FIND R	CAPITAL RECOVERY FACTOR GIVEN P TO FIND R	COMPOUND AMOUNT FACTOR GIVEN R TO FIND S	PRESENT WORTH FACTOR GIVEN R TO FIND P	N PERIODS
	(1 + I)**N	1 / (1 + I)**N	I / ((1 + I)**N - 1)	I(1 + I)**N / ((1 + I)**N - 1)	((1 + I)**N - 1) / I	((1 + I)**N - 1) / I(1 + I)**N	
1	1.1100000	.9009009	1.0000000	1.1100000	1.0000000	.9009009	1
2	1.2321000	.8116224	.4739336	.5839336	2.1100000	1.7125233	2
3	1.3676310	.7311914	.2992131	.4092131	3.3421000	2.4437147	3
4	1.5180704	.6587310	.2123264	.3223264	4.7097310	3.1024457	4
5	1.6850582	.5934513	.1605703	.2705703	6.2278014	3.6958970	5
6	1.8704146	.5346408	.1263766	.2363766	7.9128596	4.2305379	6
7	2.0761602	.4816584	.1022153	.2122153	9.7832741	4.7121963	7
8	2.3045378	.4339265	.0843211	.1943211	11.8594343	5.1461228	8
9	2.5580369	.3909248	.0706017	.1806017	14.1639720	5.5370475	9
10	2.8394210	.3521845	.0598014	.1698014	16.7220090	5.8892320	10
11	3.1517573	.3172833	.0511210	.1611210	19.5614300	6.2065153	11
12	3.4984506	.2858408	.0440273	.1540273	22.7131872	6.4923561	12
13	3.8832802	.2575143	.0381510	.1481510	26.2116378	6.7498704	13
14	4.3104410	.2319948	.0332282	.1432282	30.0949180	6.9818652	14
15	4.7845895	.2090043	.0290652	.1390652	34.4053590	7.1908696	15

11.00%

16	5.3108943	.1882922	.0255167	.1355167	39.1899485	7.3791618	16
17	5.8950927	.1696326	.0224715	.1324715	44.5008428	7.5487944	17
18	6.5435529	.1528222	.0198429	.1298429	50.3959355	7.7016166	18
19	7.2633437	.1376776	.0175625	.1275625	56.9394884	7.8392942	19
20	8.0623115	.1240339	.0155756	.1255756	64.2028321	7.9633281	20
21	8.9491658	.1117423	.0138379	.1238379	72.2651437	8.0750704	21
22	9.9335740	.1006687	.0123131	.1223131	81.2143095	8.1757391	22
23	11.0262672	.0906925	.0109712	.1209712	91.1478835	8.2664316	23
24	12.2391566	.0817050	.0097872	.1197872	102.1741507	8.3481366	24
25	13.5854638	.0736081	.0087402	.1187402	114.4133073	8.4217447	25
26	15.0798648	.0663136	.0078126	.1178126	127.9987711	8.4880583	26
27	16.7386500	.0597420	.0069892	.1169892	143.0786359	8.5478002	27
28	18.5799014	.0538216	.0062571	.1162571	159.8172859	8.6016218	28
29	20.6236906	.0484879	.0056055	.1156055	178.3971873	8.6501098	29
30	22.8922966	.0436828	.0050246	.1150246	199.0208779	8.6937926	30
31	25.4104492	.0393539	.0045063	.1145063	221.9131745	8.7331465	31
32	28.2055986	.0354540	.0040433	.1140433	247.3236237	8.7686004	32
33	31.3082145	.0319405	.0036294	.1136294	275.5292223	8.8005409	33
34	34.7521180	.0287752	.0032591	.1132591	306.8374368	8.8293161	34
35	38.5748510	.0259236	.0029275	.1129275	341.5895548	8.8552398	35
36	42.8180846	.0233546	.0026304	.1126304	380.1644058	8.8785944	36
37	47.5280740	.0210402	.0023642	.1123642	422.9824905	8.8996346	37
38	52.7561621	.0189551	.0021254	.1121254	470.5105644	8.9185897	38
39	58.5593399	.0170767	.0019111	.1119111	523.2667265	8.9356664	39
40	65.0008673	.0153844	.0017187	.1117187	581.8260664	8.9510508	40

NOTE- **N IS EXFCNENT N

12.00 PERCENT COMPOUND INTEREST FACTORS

N PERIODS	SINGLE PAYMENT COMPOUND AMOUNT FACTOR GIVEN P TO FIND S	SINGLE PAYMENT PRESENT WORTH FACTOR GIVEN S TO FIND P	UNIFORM ANNUAL SERIES SINKING FUND FACTOR GIVEN S TO FIND R	UNIFORM ANNUAL SERIES CAPITAL RECOVERY FACTOR GIVEN P TO FIND R	UNIFORM ANNUAL SERIES COMPOUND AMOUNT FACTOR GIVEN R TO FIND S	UNIFORM ANNUAL SERIES PRESENT WORTH FACTOR GIVEN R TO FIND P	N PERIODS
	(1 + I)**N	1 / (1 + I)**N	I / ((1 + I)**N - 1)	I(1 + I)**N / ((1 + I)**N - 1)	((1 + I)**N - 1) / I	((1 + I)**N - 1) / I(1 + I)**N	
1	1.1200000	.8928571	1.0000000	1.1200000	1.0000000	.8928571	1
2	1.2544000	.7971939	.4716981	.5916981	2.1200000	1.6900510	2
3	1.4049280	.7117802	.2963490	.4163490	3.3744000	2.4018313	3
4	1.5735194	.6355181	.2092344	.3292344	4.7793280	3.0373493	4
5	1.7623417	.5674269	.1574097	.2774097	6.3528474	3.6047762	5
6	1.9738227	.5066311	.1232257	.2432257	8.1151890	4.1114073	6
7	2.2106814	.4523492	.0991177	.2191177	10.0890117	4.5637565	7
8	2.4759632	.4038832	.0813028	.2013028	12.2996931	4.9676398	8
9	2.7730788	.3606100	.0676789	.1876789	14.7756563	5.3282498	9
10	3.1058482	.3219732	.0569842	.1769842	17.5487351	5.6502230	10
11	3.4785500	.2874761	.0484154	.1684154	20.6545833	5.9376991	11
12	3.8959760	.2566751	.0414368	.1614368	24.1331333	6.1943742	12
13	4.3634931	.2291742	.0356772	.1556772	28.0291093	6.4235484	13
14	4.8871123	.2046198	.0308712	.1508712	32.3926024	6.6281682	14
15	5.4735658	.1826963	.0268242	.1468242	37.2797147	6.8108645	15

12.00%

16	6.1303937	.1631217	.0233900	.1433900	42.7532804	6.9739862	16
17	6.8660409	.1456443	.0204567	.1404567	48.8836741	7.1196305	17
18	7.6899658	.1300396	.0179373	.1379373	55.7497150	7.2496701	18
19	8.6127617	.1161068	.0157630	.1357630	63.4396808	7.3657769	19
20	9.6462931	.1036668	.0138788	.1338788	72.0524424	7.4694436	20
21	10.8038483	.0925596	.0122401	.1322401	81.6987355	7.5620032	21
22	12.1003101	.0826425	.0108105	.1308105	92.5025838	7.6446457	22
23	13.5523473	.0737880	.0095600	.1295600	104.6028939	7.7184337	23
24	15.1786289	.0658821	.0084634	.1284634	118.1552411	7.7843158	24
25	17.0000644	.0588233	.0075000	.1275000	133.3338701	7.8431391	25
26	19.0400721	.0525208	.0066519	.1266519	150.3339345	7.8956599	26
27	21.3248808	.0468936	.0059041	.1259041	169.3740066	7.9425535	27
28	23.8838665	.0418693	.0052439	.1252439	190.6988874	7.9844228	28
29	26.7499305	.0373833	.0046602	.1246602	214.5827539	8.0218060	29
30	29.9599221	.0333779	.0041437	.1241437	241.3326843	8.0551840	30
31	33.5551128	.0298017	.0036861	.1236861	271.2926065	8.0849857	31
32	37.5817263	.0266087	.0032803	.1232803	304.8477192	8.1115944	32
33	42.0915335	.0237577	.0029203	.1229203	342.4294455	8.1353521	33
34	47.1425175	.0212123	.0026006	.1226006	384.5209790	8.1565644	34
35	52.7996196	.0189395	.0023166	.1223166	431.6634965	8.1755039	35
36	59.1355739	.0169103	.0020641	.1220641	484.4631161	8.1924142	36
37	66.2318428	.0150985	.0018396	.1218396	543.5986900	8.2075127	37
38	74.1796639	.0134808	.0016398	.1216398	609.8305328	8.2209935	38
39	83.0812236	.0120364	.0014620	.1214620	684.0101967	8.2330299	39
40	93.0509704	.0107468	.0013036	.1213036	767.0914203	8.2437767	40

NOTE- **N IS EXPONENT N

13.00 PERCENT COMPOUND INTEREST FACTORS

N PERIODS	SINGLE PAYMENT		UNIFORM ANNUAL SERIES				N PERIODS
	COMPOUND AMOUNT FACTOR GIVEN P TO FIND S	PRESENT WORTH FACTOR GIVEN S TO FIND P	SINKING FUND FACTOR GIVEN S TO FIND R	CAPITAL RECOVERY FACTOR GIVEN P TO FIND R	COMPOUND AMOUNT FACTOR GIVEN R TO FIND S	PRESENT WORTH FACTOR GIVEN R TO FIND P	
	(1 + I)**N	1 / (1 + I)**N	I / ((1 + I)**N - 1)	I(1 + I)**N / ((1 + I)**N - 1)	((1 + I)**N - 1) / I	((1 + I)**N - 1) / I(1 + I)**N	
1	1.1300000	.8849558	1.0000000	1.1300000	1.0000000	.8849558	1
2	1.2769000	.7831467	.4694836	.5994836	2.1300000	1.6681024	2
3	1.4428970	.6930502	.2935220	.4235220	3.4069000	2.3611526	3
4	1.6304736	.6133187	.2061942	.3361942	4.8497970	2.9744713	4
5	1.8424352	.5427599	.1543145	.2843145	6.4802706	3.5172313	5
6	2.0819518	.4803185	.1201532	.2501532	8.3227058	3.9975498	6
7	2.3526055	.4250606	.0961108	.2261108	10.4046575	4.4226104	7
8	2.6584442	.3761599	.0783867	.2083867	12.7572630	4.7987703	8
9	3.0040419	.3328848	.0648689	.1948689	15.4157072	5.1316551	9
10	3.3945674	.2945883	.0542896	.1842896	18.4197492	5.4262435	10
11	3.8358612	.2606977	.0458415	.1758415	21.8143165	5.6869411	11
12	4.3345231	.2307059	.0389861	.1689861	25.6501777	5.9176470	12
13	4.8980111	.2041645	.0333503	.1633503	29.9847008	6.1218115	13
14	5.5347525	.1806766	.0286675	.1586675	34.8827119	6.3024881	14
15	6.2542704	.1598908	.0247418	.1547418	40.4174644	6.4623788	15

13.00%

16	7.0673255	.1414962	.0214262	.1514262	46.6717348	6.6038751	16
17	7.9860778	.1252179	.0186084	.1486084	53.7390603	6.7290930	17
18	9.0242680	.1108123	.0162009	.1462009	61.7251382	6.8399053	18
19	10.1974228	.0980640	.0141344	.1441344	70.7494062	6.9379693	19
20	11.5230878	.0867823	.0123538	.1423538	80.9468290	7.0247516	20
21	13.0210892	.0767985	.0108143	.1408143	92.4699167	7.1015501	21
22	14.7138308	.0679633	.0094795	.1394795	105.4910059	7.1695133	22
23	16.6266288	.0601445	.0083191	.1383191	120.2048367	7.2296578	23
24	18.7880905	.0532252	.0073083	.1373083	136.8314654	7.2828830	24
25	21.2305423	.0471020	.0064259	.1364259	155.6195559	7.3299850	25
26	23.9905128	.0416831	.0056545	.1356545	176.8500982	7.3716681	26
27	27.1092794	.0368877	.0049791	.1349791	200.8406110	7.4085559	27
28	30.6334858	.0326440	.0043869	.1343869	227.9498904	7.4411999	28
29	34.6158389	.0288885	.0038672	.1338672	258.5833762	7.4700884	29
30	39.1158980	.0255651	.0034107	.1334107	293.1992151	7.4956534	30
31	44.2009647	.0226239	.0030092	.1330092	332.3151130	7.5182774	31
32	49.9470901	.0200212	.0026559	.1326559	376.5160777	7.5382986	32
33	56.4402118	.0177179	.0023449	.1323449	426.4631678	7.5560164	33
34	63.7774394	.0156795	.0020708	.1320708	482.9033796	7.5716960	34
35	72.0685065	.0138757	.0018292	.1318292	546.6808190	7.5855716	35
36	81.4374123	.0122794	.0016162	.1316162	618.7493254	7.5978510	36
37	92.0242759	.0108667	.0014282	.1314282	700.1867377	7.6087177	37
38	103.9874318	.0096165	.0012623	.1312623	792.2110137	7.6183343	38
39	117.5057979	.0085102	.0011158	.1311158	896.1984454	7.6268445	39
40	132.7815516	.0075312	.0009865	.1309865	1013.7042433	7.6343756	40

NOTE- **N IS EXPONENT N

14.00 PERCENT COMPOUND INTEREST FACTORS

N PERIODS	SINGLE PAYMENT COMPOUND AMOUNT FACTOR GIVEN P TO FIND S	SINGLE PAYMENT PRESENT WORTH FACTOR GIVEN S TO FIND P	UNIFORM ANNUAL SERIES SINKING FUND FACTOR GIVEN S TO FIND R	UNIFORM ANNUAL SERIES CAPITAL RECOVERY FACTOR GIVEN P TO FIND R	UNIFORM ANNUAL SERIES COMPOUND AMOUNT FACTOR GIVEN R TO FIND S	UNIFORM ANNUAL SERIES PRESENT WORTH FACTOR GIVEN R TO FIND P	N PERIODS
	(1 + I)**N	1 / (1 + I)**N	I / ((1 + I)**N - 1)	I(1 + I)**N / ((1 + I)**N - 1)	((1 + I)**N - 1) / I	((1 + I)**N - 1) / I(1 + I)**N	
1	1.1400000	.8771930	1.0000000	1.1400000	1.0000000	.8771930	1
2	1.2996000	.7694675	.4672897	.6072897	2.1400000	1.6466605	2
3	1.4815440	.6749715	.2907315	.4307315	3.4396000	2.3216320	3
4	1.6889602	.5920803	.2032048	.3432048	4.9211440	2.9137123	4
5	1.9254146	.5193687	.1512835	.2912835	6.6101042	3.4330810	5
6	2.1949726	.4555865	.1171575	.2571575	8.5355187	3.8886675	6
7	2.5022688	.3996373	.0931924	.2331924	10.7304914	4.2883048	7
8	2.8525864	.3505591	.0755700	.2155700	13.2327602	4.6388639	8
9	3.2519485	.3075079	.0621684	.2021684	16.0853466	4.9463718	9
10	3.7072213	.2697438	.0517135	.1917135	19.3372951	5.2161156	10
11	4.2262323	.2366174	.0433943	.1833943	23.0445164	5.4527330	11
12	4.8179048	.2075591	.0366693	.1766693	27.2707487	5.6602921	12
13	5.4924115	.1820694	.0311637	.1711637	32.0886535	5.8423615	13
14	6.2613491	.1597100	.0266091	.1666091	37.5810650	6.0020715	14
15	7.1379380	.1400965	.0228090	.1628090	43.8424141	6.1421680	15

14.00%

16	8.1372493	.1228917	.0196154	.1596154	50.9803521	6.2650596	16
17	9.2764642	.1077997	.0169154	.1569154	59.1176014	6.3728593	17
18	10.5751692	.0945611	.0146212	.1546212	68.3940656	6.4674205	18
19	12.0556929	.0829484	.0126632	.1526632	78.9692348	6.5503688	19
20	13.7434399	.0727617	.0109860	.1509860	91.0249277	6.6231306	20
21	15.6675785	.0638261	.0095449	.1495449	104.7684175	6.6869566	21
22	17.8610394	.0559878	.0083032	.1483032	120.4359960	6.7429444	22
23	20.3615850	.0491121	.0072308	.1472308	138.2970354	6.7920565	23
24	23.2122069	.0430808	.0063028	.1463028	158.6586204	6.8351373	24
25	26.4619158	.0377902	.0054984	.1454984	181.8708272	6.8729274	25
26	30.1665840	.0331493	.0048000	.1448000	208.3327430	6.9060767	26
27	34.3899058	.0290783	.0041929	.1441929	238.4993271	6.9351550	27
28	39.2044926	.0255073	.0036645	.1436645	272.8892329	6.9606623	28
29	44.6931216	.0223748	.0032042	.1432042	312.0937255	6.9830371	29
30	50.9501586	.0196270	.0028028	.1428028	356.7868470	7.0026641	30
31	58.0831808	.0172167	.0024526	.1424526	407.7370056	7.0198808	31
32	66.2148261	.0151024	.0021468	.1421468	465.8201864	7.0349832	32
33	75.4849017	.0132477	.0018796	.1418796	532.0350125	7.0482308	33
34	86.0527880	.0116208	.0016460	.1416460	607.5199142	7.0598516	34
35	98.1001783	.0101937	.0014418	.1414418	693.5727022	7.0700453	35
36	111.8342033	.0089418	.0012631	.1412631	791.6728805	7.0789871	36
37	127.4909917	.0078437	.0011068	.1411068	903.5070838	7.0868308	37
38	145.3397306	.0068804	.0009699	.1409699	1030.9980755	7.0937112	38
39	165.6872929	.0060355	.0008501	.1408501	1176.3378061	7.0997467	39
40	188.8835139	.0052943	.0007451	.1407451	1342.0250990	7.1050409	40

NOTE- **N IS EXPONENT N

15.00 PERCENT COMPOUND INTEREST FACTORS

N PERIODS	SINGLE PAYMENT: COMPOUND AMOUNT FACTOR GIVEN P TO FIND S (1 + I)**N	SINGLE PAYMENT: PRESENT WORTH FACTOR GIVEN S TO FIND P 1 / (1 + I)**N	UNIFORM ANNUAL SERIES: SINKING FUND FACTOR GIVEN S TO FIND R I / ((1 + I)**N - 1)	UNIFORM ANNUAL SERIES: CAPITAL RECOVERY FACTOR GIVEN P TO FIND R I(1 + I)**N / ((1 + I)**N - 1)	UNIFORM ANNUAL SERIES: COMPOUND AMOUNT FACTOR GIVEN R TO FIND S ((1 + I)**N - 1) / I	UNIFORM ANNUAL SERIES: PRESENT WORTH FACTOR GIVEN R TO FIND P ((1 + I)**N - 1) / I(1 + I)**N	N PERIODS
1	1.1500000	.8695652	1.0000000	1.1500000	1.0000000	.8695652	1
2	1.3225000	.7561437	.4651163	.6151163	2.1500000	1.6257089	2
3	1.5208750	.6575162	.2879770	.4379770	3.4725000	2.2832251	3
4	1.7490063	.5717532	.2002654	.3502654	4.9933750	2.8549784	4
5	2.0113572	.4971767	.1483156	.2983156	6.7423813	3.3521551	5
6	2.3130608	.4323276	.1142369	.2642369	8.7537384	3.7844827	6
7	2.6600199	.3759370	.0903604	.2403604	11.0667992	4.1604197	7
8	3.0590229	.3269018	.0728501	.2228501	13.7268191	4.4873215	8
9	3.5178763	.2842624	.0595740	.2095740	16.7858419	4.7715839	9
10	4.0455577	.2471847	.0492521	.1992521	20.3037182	5.0187686	10
11	4.6523914	.2149432	.0410690	.1910690	24.3492760	5.2337118	11
12	5.3502501	.1869072	.0344808	.1844808	29.0016674	5.4206190	12
13	6.1527876	.1625280	.0291105	.1791105	34.3519175	5.5831470	13
14	7.0757058	.1413287	.0246885	.1746885	40.5047051	5.7244756	14
15	8.1370616	.1228945	.0210171	.1710171	47.5804109	5.8473701	15
16	9.3576209	.1068648	.0179477	.1679477	55.7174725	5.9542349	16

15.00%

17	10.7612640	.0929259	.0153669	.1653669	65.0750934	6.0471608	17
18	12.3754536	.0808051	.0131863	.1631863	75.8363574	6.1279659	18
19	14.2317716	.0702653	.0113364	.1613364	88.2118110	6.1982312	19
20	16.3665374	.0611003	.0097615	.1597615	102.4435826	6.2593315	20
21	18.8215180	.0531307	.0084168	.1584168	118.8101200	6.3124622	21
22	21.6447457	.0462006	.0072658	.1572658	137.6316380	6.3586627	22
23	24.8914576	.0401744	.0062784	.1562784	159.2763837	6.3988372	23
24	28.6251762	.0349343	.0054298	.1554298	184.1678413	6.4337714	24
25	32.9189526	.0303776	.0046994	.1546994	212.7930175	6.4641491	25
26	37.8567955	.0264153	.0040698	.1540698	245.7119701	6.4905644	26
27	43.5353148	.0229699	.0035265	.1535265	283.5687656	6.5135343	27
28	50.0656121	.0199738	.0030571	.1530571	327.1040804	6.5335081	28
29	57.5754539	.0173685	.0026513	.1526513	377.1696925	6.5508766	29
30	66.2117720	.0151031	.0023002	.1523002	434.7451464	6.5659796	30
31	76.1435378	.0131331	.0019962	.1519962	500.9569183	6.5791127	31
32	87.5650684	.0114201	.0017328	.1517328	577.1004561	6.5905328	32
33	100.6998287	.0099305	.0015045	.1515045	664.6655245	6.6004633	33
34	115.8048030	.0086352	.0013066	.1513066	765.3653532	6.6090985	34
35	133.1755234	.0075089	.0011349	.1511349	881.1701561	6.6166074	35
36	153.1518519	.0065295	.0009859	.1509859	1014.3456796	6.6231369	36
37	176.1246297	.0056778	.0008565	.1508565	1167.4975315	6.6288147	37
38	202.5433242	.0049372	.0007443	.1507443	1343.6221612	6.6337519	38
39	232.9243228	.0042932	.0006468	.1506468	1546.1654854	6.6380451	39
40	267.8635462	.0037332	.0005621	.1505621	1779.0903082	6.6417784	40

NOTE- **N IS EXPONENT N

16.00 PERCENT COMPOUND INTEREST FACTORS

	SINGLE PAYMENT		UNIFORM ANNUAL SERIES				
N PERIODS	COMPOUND AMOUNT FACTOR GIVEN P TO FIND S	PRESENT WORTH FACTOR GIVEN S TO FIND P	SINKING FUND FACTOR GIVEN S TO FIND R	CAPITAL RECOVERY FACTOR GIVEN P TO FIND R	COMPOUND AMOUNT FACTOR GIVEN R TO FIND S	PRESENT WORTH FACTOR GIVEN R TO FIND P	N PERIODS
	(1 + I)**N	1 / (1 + I)**N	I / ((1 + I)**N - 1)	I(1 + I)**N / ((1 + I)**N - 1)	((1 + I)**N - 1) / I	((1 + I)**N - 1) / (I(1 + I)**N)	
1	1.1600000	.8620690	1.0000000	1.1600000	1.0000000	.8620690	1
2	1.3456000	.7431629	.4629630	.6229630	2.1600000	1.6052319	2
3	1.5608960	.6406577	.2852579	.4452579	3.5056000	2.2458895	3
4	1.8106394	.5522911	.1973751	.3573751	5.0664960	2.7981806	4
5	2.1003417	.4761130	.1454094	.3054094	6.8771354	3.2742937	5
6	2.4363963	.4104423	.1113899	.2713899	8.9774770	3.6847359	6
7	2.8262197	.3538295	.0876127	.2476127	11.4138733	4.0385654	7
8	3.2784149	.3050255	.0702243	.2302243	14.2400931	4.3435909	8
9	3.8029613	.2629530	.0570825	.2170825	17.5185080	4.6065439	9
10	4.4114351	.2266836	.0469011	.2069011	21.3214692	4.8332275	10
11	5.1172647	.1954169	.0388608	.1988608	25.7329043	5.0286444	11
12	5.9360270	.1684628	.0324147	.1924147	30.8501690	5.1971072	12
13	6.8857914	.1452266	.0271841	.1871841	36.7861961	5.3423338	13
14	7.9875180	.1251953	.0228980	.1828980	43.6719874	5.4675291	14
15	9.2655209	.1079270	.0193575	.1793575	51.6595054	5.5754562	15

16.00%

16	10.7480042	.0930405	.0164136	.1764136	60.9250263	5.6684967	16
17	12.4676849	.0802074	.0139522	.1739522	71.6730305	5.7487040	17
18	14.4625145	.0691443	.0118849	.1718849	84.1407154	5.8178483	18
19	16.7765168	.0596071	.0101417	.1701417	98.6032298	5.8774554	19
20	19.4607595	.0513855	.0086670	.1686670	115.3797466	5.9288409	20
21	22.5744810	.0442978	.0074162	.1674162	134.8405060	5.9731387	21
22	26.1863979	.0381878	.0063526	.1663526	157.4149870	6.0113265	22
23	30.3762216	.0329205	.0054466	.1654466	183.6013849	6.0442470	23
24	35.2364170	.0283797	.0046734	.1646734	213.9776065	6.0726267	24
25	40.8742438	.0244653	.0040126	.1640126	249.2140235	6.0970920	25
26	47.4141228	.0210908	.0034472	.1634472	290.0882673	6.1181827	26
27	55.0003824	.0181817	.0029629	.1629629	337.5023901	6.1363644	27
28	63.8004436	.0156739	.0025478	.1625478	392.5027725	6.1520383	28
29	74.0085146	.0135120	.0021915	.1621915	456.3032161	6.1655503	29
30	85.8498769	.0116482	.0018857	.1618857	530.3117307	6.1771985	30
31	99.5858572	.0100416	.0016230	.1616230	616.1616076	6.1872401	31
32	115.5195944	.0086565	.0013971	.1613971	715.7474648	6.1958966	32
33	134.0027295	.0074625	.0012030	.1612030	831.2670592	6.2033592	33
34	155.4431662	.0064332	.0010360	.1610360	965.2697886	6.2097924	34
35	180.3140728	.0055459	.0008923	.1608923	1120.7129548	6.2153383	35
36	209.1643244	.0047809	.0007686	.1607686	1301.0270276	6.2201192	36
37	242.6306163	.0041215	.0006622	.1606622	1510.1913520	6.2242407	37
38	281.4515149	.0035530	.0005705	.1605705	1752.8219683	6.2277937	38
39	326.4837573	.0030629	.0004916	.1604916	2034.2734833	6.2308566	39
40	378.7211585	.0026405	.0004236	.1604236	2360.7572406	6.2334971	40

NOTE- **N IS EXPONENT N

17.00 PERCENT COMPOUND INTEREST FACTORS

N PERIODS	SINGLE PAYMENT: COMPOUND AMOUNT FACTOR GIVEN P TO FIND S	SINGLE PAYMENT: PRESENT WORTH FACTOR GIVEN S TO FIND P	UNIFORM ANNUAL SERIES: SINKING FUND FACTOR GIVEN S TO FIND R	UNIFORM ANNUAL SERIES: CAPITAL RECOVERY FACTOR GIVEN P TO FIND R	UNIFORM ANNUAL SERIES: COMPOUND AMOUNT FACTOR GIVEN R TO FIND S	UNIFORM ANNUAL SERIES: PRESENT WORTH FACTOR GIVEN R TO FIND P	N PERIODS
	(1 + I)**N	1 / (1 + I)**N	I / ((1 + I)**N - 1)	I(1 + I)**N / ((1 + I)**N - 1)	((1 + I)**N - 1) / I	((1 + I)**N - 1) / I(1 + I)**N	
1	1.1700000	.8547009	1.0000000	1.1700000	1.0000000	.8547009	1
2	1.3689000	.7305136	.4608295	.6308295	2.1700000	1.5852144	2
3	1.6016130	.6243706	.2825737	.4525737	3.5389000	2.2095850	3
4	1.8738872	.5336500	.1945331	.3645331	5.1405130	2.7432350	4
5	2.1924480	.4561112	.1425639	.3125639	7.0144002	3.1993462	5
6	2.5651642	.3898386	.1086148	.2786148	9.2068482	3.5891848	6
7	3.0012421	.3331954	.0849472	.2549472	11.7720124	3.9223801	7
8	3.5114533	.2847824	.0676899	.2376899	14.7732546	4.2071625	8
9	4.1084003	.2434037	.0546905	.2246905	18.2847078	4.4505662	9
10	4.8068284	.2080374	.0446566	.2146566	22.3931082	4.6586036	10
11	5.6239892	.1778097	.0367648	.2067648	27.1999366	4.8364134	11
12	6.5800674	.1519741	.0304656	.2004656	32.8239258	4.9883875	12
13	7.6986788	.1298924	.0253781	.1953781	39.4039932	5.1182799	13
14	9.0074542	.1110192	.0212302	.1912302	47.1026720	5.2292991	14
15	10.5387215	.0948882	.0178221	.1878221	56.1101262	5.3241872	15

17.00%

16	12.3303041	.0811010	.0150040	.1850040	66.6488477	5.4052882	16
17	14.4264558	.0693171	.0126616	.1826616	78.9791518	5.4746053	17
18	16.8789533	.0592454	.0107060	.1807060	93.4056076	5.5338507	18
19	19.7483754	.0506371	.0090675	.1790675	110.2845609	5.5844878	19
20	23.1055992	.0432796	.0076904	.1776904	130.0329363	5.6277673	20
21	27.0335510	.0369911	.0065300	.1765300	153.1385354	5.6647584	21
22	31.6292547	.0316163	.0055502	.1755502	180.1720864	5.6963747	22
23	37.0062280	.0270225	.0047214	.1747214	211.8013411	5.7233972	23
24	43.2972868	.0230961	.0040192	.1740192	248.8075691	5.7464933	24
25	50.6578255	.0197403	.0034234	.1734234	292.1048559	5.7662336	25
26	59.2696558	.0168720	.0029175	.1729175	342.7626814	5.7831056	26
27	69.3454973	.0144205	.0024874	.1724874	402.0323372	5.7975262	27
28	81.1342319	.0123253	.0021214	.1721214	471.3778345	5.8098514	28
29	94.9270513	.0105344	.0018099	.1718099	552.5120664	5.8203859	29
30	111.0646500	.0090038	.0015445	.1715445	647.4391177	5.8293896	30
31	129.9456405	.0076955	.0013184	.1713184	758.5037677	5.8370851	31
32	152.0363994	.0065774	.0011256	.1711256	888.4494082	5.8436625	32
33	177.8825873	.0056217	.0009611	.1709611	1040.4358076	5.8492842	33
34	208.1226271	.0048049	.0008208	.1708208	1218.3683949	5.8540891	34
35	243.5034738	.0041067	.0007010	.1707010	1426.4910221	5.8581958	35
36	284.8990643	.0035100	.0005988	.1705988	1669.9944958	5.8617058	36
37	333.3319052	.0030000	.0005115	.1705115	1954.8935601	5.8647058	37
38	389.9983291	.0025641	.0004370	.1704370	2288.2254653	5.8672699	38
39	456.2980451	.0021916	.0003734	.1703734	2678.2237944	5.8694615	39
40	533.8687127	.0018731	.0003190	.1703190	3134.5218395	5.8713346	40

NOTE- **N IS EXPONENT N

18.00 PERCENT COMPOUND INTEREST FACTORS

N PERIODS	SINGLE PAYMENT: COMPOUND AMOUNT FACTOR GIVEN P TO FIND S	SINGLE PAYMENT: PRESENT WORTH FACTOR GIVEN S TO FIND P	UNIFORM ANNUAL SERIES: SINKING FUND FACTOR GIVEN S TO FIND R	UNIFORM ANNUAL SERIES: CAPITAL RECOVERY FACTOR GIVEN P TO FIND R	UNIFORM ANNUAL SERIES: COMPOUND AMOUNT FACTOR GIVEN R TO FIND S	UNIFORM ANNUAL SERIES: PRESENT WORTH FACTOR GIVEN R TO FIND P	N PERIODS
	(1 + I)**N	1 / (1 + I)**N	I / ((1 + I)**N - 1)	I(1 + I)**N / ((1 + I)**N - 1)	((1 + I)**N - 1) / I	((1 + I)**N - 1) / (I(1 + I)**N)	
1	1.1800000	.8474576	1.0000000	1.1800000	1.0000000	.8474576	1
2	1.3924000	.7181844	.4587156	.6387156	2.1800000	1.5656421	2
3	1.6430320	.6086309	.2799239	.4599239	3.5724000	2.1742729	3
4	1.9387778	.5157889	.1917387	.3717387	5.2154320	2.6900618	4
5	2.2877578	.4371092	.1397778	.3197778	7.1542098	3.1271710	5
6	2.6995542	.3704315	.1059101	.2859101	9.4419675	3.4976026	6
7	3.1854739	.3139250	.0823620	.2623620	12.1415217	3.8115276	7
8	3.7588592	.2660382	.0652444	.2452444	15.3269956	4.0775658	8
9	4.4354539	.2254561	.0523948	.2323948	19.0858548	4.3030218	9
10	5.2338356	.1910645	.0425146	.2225146	23.5213086	4.4940863	10
11	6.1759260	.1619190	.0347764	.2147764	28.7551442	4.6560053	11
12	7.2875926	.1372195	.0286278	.2086278	34.9310701	4.7932249	12
13	8.5993593	.1162877	.0236862	.2036862	42.2186628	4.9095126	13
14	10.1472440	.0985489	.0196781	.1996781	50.8180221	5.0080615	14
15	11.9737479	.0835160	.0164028	.1964028	60.9652660	5.0915776	15

18.00%

16	14.1290225	.0707763	.0137101	.1937101	72.9390139	5.1623539	16
17	16.6722466	.0599799	.0114853	.1914853	87.0680364	5.2223338	17
18	19.6732509	.0508304	.0096395	.1896395	103.7402830	5.2731642	18
19	23.2144361	.0430766	.0081028	.1881028	123.4135339	5.3162409	19
20	27.3930346	.0365056	.0068200	.1868200	146.6279700	5.3527465	20
21	32.3237808	.0309370	.0057464	.1857464	174.0210046	5.3836835	21
22	38.1420614	.0262178	.0048463	.1848463	206.3447855	5.4099012	22
23	45.0076324	.0222185	.0040902	.1840902	244.4868468	5.4321197	23
24	53.1090063	.0188292	.0034543	.1834543	289.4944793	5.4509489	24
25	62.6686274	.0159569	.0029188	.1829188	342.6034855	5.4669058	25
26	73.9489803	.0135228	.0024675	.1824675	405.2721129	5.4804287	26
27	87.2597968	.0114600	.0020867	.1820867	479.2210933	5.4918887	27
28	102.9665602	.0097119	.0017653	.1817653	566.4808901	5.5016006	28
29	121.5005410	.0082304	.0014938	.1814938	669.4474503	5.5098310	29
30	143.3706384	.0069749	.0012643	.1812643	790.9479913	5.5168060	30
31	169.1773534	.0059110	.0010703	.1810703	934.3186298	5.5227169	31
32	199.6292770	.0050093	.0009062	.1809062	1103.4959831	5.5277262	32
33	235.5625468	.0042452	.0007674	.1807674	1303.1252601	5.5319713	33
34	277.9638052	.0035976	.0006499	.1806499	1538.6878069	5.5355689	34
35	327.9972902	.0030488	.0005505	.1805505	1816.6516121	5.5386177	35
36	387.0368024	.0025837	.0004663	.1804663	2144.6489023	5.5412015	36
37	456.7034269	.0021896	.0003950	.1803950	2531.6857047	5.5433911	37
38	538.9100437	.0018556	.0003346	.1803346	2988.3891316	5.5452467	38
39	635.9138515	.0015725	.0002835	.1802835	3527.2991753	5.5468192	39
40	750.3783448	.0013327	.0002402	.1802402	4163.2130268	5.5481519	40

NOTE- **N IS EXPONENT N

19.00 PERCENT COMPOUND INTEREST FACTORS

N PERIODS	SINGLE PAYMENT: COMPOUND AMOUNT FACTOR GIVEN P TO FIND S	SINGLE PAYMENT: PRESENT WORTH FACTOR GIVEN S TO FIND P	UNIFORM ANNUAL SERIES: SINKING FUND FACTOR GIVEN S TO FIND R	UNIFORM ANNUAL SERIES: CAPITAL RECOVERY FACTOR GIVEN P TO FIND R	UNIFORM ANNUAL SERIES: COMPOUND AMOUNT FACTOR GIVEN R TO FIND S	UNIFORM ANNUAL SERIES: PRESENT WORTH FACTOR GIVEN R TO FIND P	N PERIODS
	(1 + I)**N	1 / (1 + I)**N	I / ((1 + I)**N - 1)	I(1 + I)**N / ((1 + I)**N - 1)	((1 + I)**N - 1) / I	((1 + I)**N - 1) / (I(1 + I)**N)	
1	1.1900000	.8403361	1.0000000	1.1900000	1.0000000	.8403361	1
2	1.4161000	.7061648	.4566210	.6466210	2.1900000	1.5465010	2
3	1.6851590	.5934158	.2773079	.4673079	3.6061000	2.1399168	3
4	2.0053392	.4986688	.1889909	.3789909	5.2912590	2.6385855	4
5	2.3863537	.4190494	.1370502	.3270502	7.2965982	3.0576349	5
6	2.8397609	.3521423	.1032743	.2932743	9.6829519	3.4097772	6
7	3.3793154	.2959179	.0798549	.2698549	12.5227127	3.7056951	7
8	4.0213853	.2486705	.0628851	.2528851	15.9020281	3.9543657	8
9	4.7854486	.2089668	.0501922	.2401922	19.9234135	4.1633325	9
10	5.6946838	.1756024	.0404713	.2304713	24.7088621	4.3389349	10
11	6.7766737	.1475650	.0328909	.2228909	30.4035458	4.4864999	11
12	8.0642417	.1240042	.0268960	.2168960	37.1802196	4.6105041	12
13	9.5964476	.1042052	.0221022	.2121022	45.2444613	4.7147093	13
14	11.4197727	.0875674	.0182346	.2082346	54.8409089	4.8022768	14
15	13.5895295	.0735861	.0150919	.2050919	66.2606816	4.8758628	15

19.00%

16	16.1715401	.0618370	.0125234	.2025234	79.8502111	4.9376998	16
17	19.2441327	.0519639	.0104143	.2004143	96.0217512	4.9896637	17
18	22.9005180	.0436671	.0086756	.1986756	115.2658839	5.0333309	18
19	27.2516164	.0366951	.0072376	.1972376	138.1664019	5.0700259	19
20	32.4294235	.0308362	.0060453	.1960453	165.4180183	5.1008621	20
21	38.5910139	.0259128	.0050544	.1950544	197.8474417	5.1267749	21
22	45.9233066	.0217754	.0042294	.1942294	236.4384557	5.1485503	22
23	54.6487348	.0182987	.0035416	.1935416	282.3617622	5.1668490	23
24	65.0319944	.0153770	.0029673	.1929673	337.0104971	5.1822261	24
25	77.3880734	.0129219	.0024873	.1924873	402.0424915	5.1951480	25
26	92.0918073	.0108587	.0020858	.1920858	479.4305649	5.2060067	26
27	109.5892507	.0091250	.0017497	.1917497	571.5223722	5.2151317	27
28	130.4112084	.0076681	.0014682	.1914682	681.1116229	5.2227997	28
29	155.1893379	.0064437	.0012323	.1912323	811.5228313	5.2292435	29
30	184.6753122	.0054149	.0010344	.1910344	966.7121692	5.2346584	30
31	219.7636215	.0045503	.0008685	.1908685	1151.3874814	5.2392087	31
32	261.5187095	.0038238	.0007293	.1907293	1371.1511029	5.2430325	32
33	311.2072644	.0032133	.0006125	.1906125	1632.6698124	5.2462458	33
34	370.3366446	.0027002	.0005144	.1905144	1943.8770767	5.2489461	34
35	440.7006071	.0022691	.0004321	.1904321	2314.2137213	5.2512152	35
36	524.4337224	.0019068	.0003630	.1903630	2754.9143284	5.2531220	36
37	624.0761296	.0016024	.0003049	.1903049	3279.3480508	5.2547244	37
38	742.6505943	.0013465	.0002562	.1902562	3903.4241804	5.2560709	38
39	883.7542072	.0011315	.0002152	.1902152	4646.0747747	5.2572024	39
40	1051.6675066	.0009509	.0001803	.1901808	5529.8289819	5.2581533	40

NOTE- **N IS EXPONENT N

20.00 PERCENT COMPOUND INTEREST FACTORS

N PERIODS	SINGLE PAYMENT: COMPOUND AMOUNT FACTOR GIVEN P TO FIND S	SINGLE PAYMENT: PRESENT WORTH FACTOR GIVEN S TO FIND P	UNIFORM ANNUAL SERIES: SINKING FUND FACTOR GIVEN S TO FIND R	UNIFORM ANNUAL SERIES: CAPITAL RECOVERY FACTOR GIVEN P TO FIND R	UNIFORM ANNUAL SERIES: COMPOUND AMOUNT FACTOR GIVEN R TO FIND S	UNIFORM ANNUAL SERIES: PRESENT WORTH FACTOR GIVEN R TO FIND P	N PERIODS
	(1 + I)**N	1 / (1 + I)**N	I / ((1 + I)**N - 1)	I(1 + I)**N / ((1 + I)**N - 1)	((1 + I)**N - 1) / I	((1 + I)**N - 1) / (I(1 + I)**N)	
1	1.2000000	.8333333	1.0000000	1.2000000	1.0000000	.8333333	1
2	1.4400000	.6944444	.4545455	.6545455	2.2000000	1.5277778	2
3	1.7280000	.5787037	.2747253	.4747253	3.6400000	2.1064815	3
4	2.0736000	.4822531	.1862891	.3862891	5.3680000	2.5887346	4
5	2.4883200	.4018776	.1343797	.3343797	7.4416000	2.9906121	5
6	2.9859840	.3348980	.1007057	.3007057	9.9299200	3.3255101	6
7	3.5831808	.2790816	.0774239	.2774239	12.9159040	3.6045918	7
8	4.2998170	.2325680	.0606094	.2606094	16.4990848	3.8371598	8
9	5.1597804	.1938067	.0480795	.2480795	20.7989018	4.0309665	9
10	6.1917364	.1615056	.0385228	.2385228	25.9586821	4.1924721	10
11	7.4300837	.1345880	.0311038	.2311038	32.1504185	4.3270601	11
12	8.9161004	.1121567	.0252650	.2252650	39.5805022	4.4392167	12
13	10.6993205	.0934639	.0206200	.2206200	48.4966027	4.5326806	13
14	12.8391846	.0778866	.0168931	.2168931	59.1959232	4.6105672	14
15	15.4070216	.0649055	.0138821	.2138821	72.0351079	4.6754726	15

20.00%

16	18.4884259	.0540879	.0114361	.2114361	87.4421294	4.7295605	16
17	22.1861111	.0450732	.0094401	.2094401	105.9305553	4.7746338	17
18	26.6233333	.0375610	.0078054	.2078054	128.1166664	4.8121948	18
19	31.9479999	.0313009	.0064625	.2064625	154.7399997	4.8434957	19
20	38.3375999	.0260841	.0053565	.2053565	186.6879996	4.8695797	20
21	46.0051199	.0217367	.0044439	.2044439	225.0255995	4.8913164	21
22	55.2061439	.0181139	.0036896	.2036896	271.0307195	4.9094304	22
23	66.2473727	.0150949	.0030653	.2030653	326.2368633	4.9245253	23
24	79.4968472	.0125791	.0025479	.2025479	392.4842360	4.9371044	24
25	95.3962166	.0104826	.0021187	.2021187	471.9810832	4.9475870	25
26	114.4754600	.0087355	.0017625	.2017625	567.3772999	4.9563225	26
27	137.3705520	.0072796	.0014666	.2014666	681.8527598	4.9636021	27
28	164.8446624	.0060663	.0012207	.2012207	819.2233118	4.9696684	28
29	197.8135948	.0050553	.0010162	.2010162	984.0679742	4.9747237	29
30	237.3763138	.0042127	.0008461	.2008461	1181.8815690	4.9789364	30
31	284.8515766	.0035106	.0007046	.2007046	1419.2578828	4.9824470	31
32	341.8218919	.0029255	.0005868	.2005868	1704.1094594	4.9853725	32
33	410.1862702	.0024379	.0004888	.2004888	2045.9313512	4.9878104	33
34	492.2235243	.0020316	.0004071	.2004071	2456.1176215	4.9898420	34
35	590.6682292	.0016930	.0003392	.2003392	2948.3411458	4.9915350	35
36	708.8018750	.0014108	.0002826	.2002826	3539.0093749	4.9929458	36
37	850.5622500	.0011757	.0002354	.2002354	4247.8112499	4.9941215	37
38	1020.6747000	.0009797	.0001961	.2001961	5098.3734999	4.9951013	38
39	1224.8096400	.0008165	.0001634	.2001634	6119.0481999	4.9959177	39
40	1469.7715680	.0006804	.0001362	.2001362	7343.8578398	4.9965981	40

NOTE- **N IS EXPONENT N

21.00 PERCENT COMPOUND INTEREST FACTORS

	SINGLE PAYMENT		UNIFORM ANNUAL SERIES				
N PERIODS	COMPOUND AMOUNT FACTOR GIVEN P TO FIND S	PRESENT WORTH FACTOR GIVEN S TO FIND P	SINKING FUND FACTOR GIVEN S TO FIND R	CAPITAL RECOVERY FACTOR GIVEN P TO FIND R	COMPOUND AMOUNT FACTOR GIVEN R TO FIND S	PRESENT WORTH FACTOR GIVEN R TO FIND P	N PERIODS
	(1 + I)**N	1 / (1 + I)**N	I / ((1 + I)**N - 1)	I(1 + I)**N / ((1 + I)**N - 1)	((1 + I)**N - 1) / I	((1 + I)**N - 1) / (I(1 + I)**N)	
1	1.2100000	.8264463	1.0000000	1.2100000	1.0000000	.8264463	1
2	1.4641000	.6830135	.4524887	.6624887	2.2100000	1.5094597	2
3	1.7715610	.5644739	.2721755	.4821755	3.6741000	2.0739337	3
4	2.1435888	.4665074	.1836324	.3936324	5.4456610	2.5404410	4
5	2.5937425	.3855433	.1317653	.3417653	7.5892498	2.9259843	5
6	3.1384284	.3186308	.0982030	.3082030	10.1829923	3.2446152	6
7	3.7974983	.2633313	.0750671	.2850671	13.3214206	3.5079464	7
8	4.5949730	.2176291	.0584149	.2684149	17.1189190	3.7255755	8
9	5.5599173	.1798588	.0460535	.2560535	21.7138920	3.9054343	9
10	6.7274999	.1486436	.0366652	.2466652	27.2738093	4.0540780	10
11	8.1402749	.1228460	.0294106	.2394106	34.0013092	4.1769239	11
12	9.8497327	.1015256	.0237295	.2337295	42.1415842	4.2784495	12
13	11.9181765	.0839055	.0192340	.2292340	51.9913168	4.3623550	13
14	14.4209936	.0693433	.0156471	.2256471	63.9094934	4.4316983	14
15	17.4494023	.0573086	.0127664	.2227664	78.3304870	4.4890069	15

21.00%

16	21.1137767	.0473624	.0104405	.2204406	95.7798893	4.5363693	16
17	25.5476699	.0391425	.0085548	.2185548	116.8936660	4.5755118	17
18	30.9126805	.0323492	.0070204	.2170204	142.4413359	4.6078610	18
19	37.4043434	.0267349	.0057685	.2157685	173.3540164	4.6345959	19
20	45.2592556	.0220949	.0047448	.2147448	210.7583598	4.6566908	20
21	54.7636992	.0182603	.0039060	.2139060	256.0176154	4.6749511	21
22	66.2640761	.0150911	.0032177	.2132177	310.7813147	4.6900422	22
23	80.1795321	.0124720	.0026522	.2126522	377.0453907	4.7025142	23
24	97.0172338	.0103074	.0021871	.2121871	457.2249228	4.7128217	24
25	117.3908529	.0085186	.0018043	.2118043	554.2421566	4.7213402	25
26	142.0429320	.0070401	.0014889	.2114889	671.6330094	4.7283804	26
27	171.8719477	.0058183	.0012290	.2112290	813.6759414	4.7341986	27
28	207.9650567	.0048085	.0010147	.2110147	985.5478891	4.7390071	28
29	251.6377186	.0039740	.0008379	.2108379	1193.5129459	4.7429811	29
30	304.4816395	.0032843	.0006920	.2106920	1445.1506645	4.7462654	30
31	368.4227838	.0027143	.0005715	.2105715	1749.6323040	4.7489797	31
32	445.7915685	.0022432	.0004721	.2104721	2118.0550879	4.7512229	32
33	539.4077978	.0018539	.0003900	.2103900	2563.8466563	4.7530767	33
34	652.6834354	.0015321	.0003222	.2103222	3103.2544541	4.7546089	34
35	789.7469568	.0012662	.0002662	.2102662	3755.9378895	4.7558751	35
36	955.5938177	.0010465	.0002200	.2102200	4545.6848463	4.7569216	36
37	1156.2685195	.0008649	.0001818	.2101818	5501.2786640	4.7577864	37
38	1399.0849085	.0007148	.0001502	.2101502	6657.5471835	4.7585012	38
39	1692.8927393	.0005907	.0001241	.2101241	8056.6320920	4.7590919	39
40	2048.4002146	.0004882	.0001026	.2101026	9749.5248314	4.7595801	40

NOTE- **N IS EXPONENT N

22.00 PERCENT COMPOUND INTEREST FACTORS

	SINGLE PAYMENT		UNIFORM ANNUAL SERIES				
N PERIODS	COMPOUND AMOUNT FACTOR GIVEN P TO FIND S	PRESENT WORTH FACTOR GIVEN S TO FIND P	SINKING FUND FACTOR GIVEN S TO FIND R	CAPITAL RECOVERY FACTOR GIVEN P TO FIND R	COMPOUND AMOUNT FACTOR GIVEN R TO FIND S	PRESENT WORTH FACTOR GIVEN R TO FIND P	N PERIODS
	(1 + I)**N	1 / (1 + I)**N	I / ((1 + I)**N - 1)	I(1 + I)**N / ((1 + I)**N - 1)	((1 + I)**N - 1) / I	((1 + I)**N - 1) / I(1 + I)**N	
1	1.2200000	.8196721	1.0000000	1.2200000	1.0000000	.8196721	1
2	1.4884000	.6718624	.4504505	.6704505	2.2200000	1.4915345	2
3	1.8158480	.5507069	.2696581	.4896581	3.7084000	2.0422414	3
4	2.2153346	.4513991	.1810201	.4010201	5.5242480	2.4936405	4
5	2.7027082	.3699993	.1292059	.3492059	7.7395826	2.8636398	5
6	3.2973040	.3032781	.0957644	.3157644	10.4422907	3.1669178	6
7	4.0227108	.2485886	.0727824	.2927824	13.7395947	3.4155064	7
8	4.9077072	.2037611	.0562990	.2762990	17.7623055	3.6192676	8
9	5.9874028	.1670173	.0441111	.2641111	22.6700127	3.7862849	9
10	7.3046314	.1368994	.0348950	.2548950	28.6574155	3.9231843	10
11	8.9116503	.1122127	.0278071	.2478071	35.9620469	4.0353970	11
12	10.8722134	.0919776	.0222848	.2422848	44.8736973	4.1273746	12
13	13.2641003	.0753915	.0179385	.2379385	55.7459107	4.2027661	13
14	16.1822024	.0617963	.0144907	.2344907	69.0100110	4.2645623	14
15	19.7422870	.0506527	.0117382	.2317382	85.1922134	4.3152150	15

22.00%

16	24.0855901	.0415186	.0095298	.2295298	104.9345004	4.3567336	16
17	29.3844199	.0340316	.0077507	.2277507	129.0200905	4.3907653	17
18	35.8489923	.0278948	.0063130	.2263130	158.4045104	4.4186601	18
19	43.7357706	.0228646	.0051479	.2251479	194.2535027	4.4415246	19
20	53.3576401	.0187415	.0042019	.2242019	237.9892733	4.4602661	20
21	65.0963209	.0153619	.0034323	.2234323	291.3469134	4.4756279	21
22	79.4175115	.0125917	.0028055	.2228055	356.4432343	4.4882196	22
23	96.8893641	.0103211	.0022943	.2222943	435.8607459	4.4985407	23
24	118.2050242	.0084599	.0018771	.2218771	532.7501099	4.5070006	24
25	144.2101295	.0069343	.0015362	.2215362	650.9551341	4.5139349	25
26	175.9363580	.0056839	.0012576	.2212576	795.1652636	4.5196188	26
27	214.6423568	.0046589	.0010298	.2210298	971.1016216	4.5242777	27
28	261.8636752	.0038188	.0008434	.2208434	1185.7439784	4.5280965	28
29	319.4736838	.0031301	.0006908	.2206908	1447.6076536	4.5312266	29
30	389.7578942	.0025657	.0005659	.2205659	1767.0813374	4.5337923	30
31	475.5046310	.0021030	.0004636	.2204636	2156.8392317	4.5358953	31
32	580.1156498	.0017238	.0003799	.2203799	2632.3438627	4.5376191	32
33	707.7410927	.0014129	.0003113	.2203113	3212.4595124	4.5390321	33
34	863.4441331	.0011582	.0002551	.2202551	3920.2006052	4.5401902	34
35	1053.4018424	.0009493	.0002090	.2202090	4783.6447383	4.5411395	35
36	1285.1502478	.0007781	.0001713	.2201713	5837.0465808	4.5419176	36
37	1567.8833023	.0006378	.0001404	.2201404	7122.1968285	4.5425554	37
38	1912.8176288	.0005228	.0001151	.2201151	8690.0801308	4.5430782	38
39	2333.6375071	.0004285	.0000943	.2200943	10602.8977596	4.5435067	39
40	2847.0377587	.0003512	.0000773	.2200773	12936.5352667	4.5438580	40

NOTE- **N IS EXPONENT N

23.00 PERCENT COMPOUND INTEREST FACTORS

N PERIODS	SINGLE PAYMENT: COMPOUND AMOUNT FACTOR GIVEN P TO FIND S	SINGLE PAYMENT: PRESENT WORTH FACTOR GIVEN S TO FIND P	UNIFORM ANNUAL SERIES: SINKING FUND FACTOR GIVEN S TO FIND R	UNIFORM ANNUAL SERIES: CAPITAL RECOVERY FACTOR GIVEN P TO FIND R	UNIFORM ANNUAL SERIES: COMPOUND AMOUNT FACTOR GIVEN R TO FIND S	UNIFORM ANNUAL SERIES: PRESENT WORTH FACTOR GIVEN R TO FIND P	N PERIODS
	(1 + I)**N	1 / (1 + I)**N	I / ((1 + I)**N - 1)	I(1 + I)**N / ((1 + I)**N - 1)	((1 + I)**N - 1) / I	((1 + I)**N - 1) / (I(1 + I)**N)	
1	1.2300000	.8130081	1.0000000	1.2300000	1.0000000	.8130081	1
2	1.5129000	.6609822	.4484305	.6784305	2.2300000	1.4739903	2
3	1.8608670	.5373839	.2671725	.4971725	3.7429000	2.0113743	3
4	2.2888664	.4368975	.1784514	.4084514	5.6037670	2.4482718	4
5	2.8153057	.3552012	.1267004	.3567004	7.8926334	2.8034730	5
6	3.4628260	.2887815	.0933887	.3233887	10.7079391	3.0922545	6
7	4.2592760	.2347817	.0705678	.3005678	14.1707651	3.3270361	7
8	5.2389094	.1908794	.0542592	.2842592	18.4300411	3.5179156	8
9	6.4438586	.1551865	.0422494	.2722494	23.6689505	3.6731021	9
10	7.9259461	.1261679	.0332085	.2632085	30.1128091	3.7992700	10
11	9.7489137	.1025755	.0262890	.2562890	38.0387552	3.9018455	11
12	11.9911638	.0833947	.0209259	.2509259	47.7876689	3.9852403	12
13	14.7491315	.0678006	.0167283	.2467283	59.7788328	4.0530409	13
14	18.1414318	.0551224	.0134178	.2434178	74.5279643	4.1081633	14
15	22.3139611	.0448150	.0107910	.2407910	92.6693961	4.1529783	15

23.00%

16	27.4461722	.0364350	.0086969	.2386969	114.9833572	4.1894132	16
17	33.7587917	.0296219	.0070210	.2370210	142.4295293	4.2190352	17
18	41.5233138	.0240829	.0056757	.2356757	176.1883211	4.2431180	18
19	51.0736760	.0195796	.0045932	.2345932	217.7116349	4.2626976	19
20	62.8206215	.0159183	.0037204	.2337204	268.7853109	4.2786159	20
21	77.2693645	.0129417	.0030156	.2330156	331.6059325	4.2915577	21
22	95.0413183	.0105217	.0024457	.2324457	408.8752969	4.3020794	22
23	116.9008215	.0085543	.0019845	.2319845	503.9166152	4.3106337	23
24	143.7880104	.0069547	.0016108	.2316108	620.8174367	4.3175883	24
25	176.8592528	.0056542	.0013079	.2313079	764.6054472	4.3232425	25
26	217.5368810	.0045969	.0010622	.2310622	941.4647000	4.3278395	26
27	267.5703636	.0037373	.0008628	.2308628	1159.0015810	4.3315768	27
28	329.1115473	.0030385	.0007010	.2307010	1426.5719447	4.3346153	28
29	404.8072031	.0024703	.0005696	.2305696	1755.6834919	4.3370856	29
30	497.9128599	.0020084	.0004629	.2304629	2160.4906951	4.3390940	30
31	612.4328176	.0016328	.0003762	.2303762	2658.4035549	4.3407268	31
32	753.2923657	.0013275	.0003057	.2303057	3270.8363726	4.3420543	32
33	926.5496098	.0010793	.0002485	.2302485	4024.1287383	4.3431336	33
34	1139.6560201	.0008775	.0002020	.2302020	4950.6783481	4.3440111	34
35	1401.7769047	.0007134	.0001642	.2301642	6090.3343681	4.3447244	35
36	1724.1855927	.0005800	.0001335	.2301335	7492.1112728	4.3453044	36
37	2120.7482791	.0004715	.0001085	.2301085	9216.2968656	4.3457759	37
38	2608.5203833	.0003834	.0000882	.2300882	11337.0451446	4.3461593	38
39	3208.4800714	.0003117	.0000717	.2300717	13945.5655279	4.3464710	39
40	3946.4304878	.0002534	.0000583	.2300583	17154.0455993	4.3467244	40

NOTE- **N IS EXPONENT N

24.00 PERCENT COMPOUND INTEREST FACTORS

N PERIODS	SINGLE PAYMENT: COMPOUND AMOUNT FACTOR GIVEN P TO FIND S	SINGLE PAYMENT: PRESENT WORTH FACTOR GIVEN S TO FIND P	UNIFORM ANNUAL SERIES: SINKING FUND FACTOR GIVEN S TO FIND R	UNIFORM ANNUAL SERIES: CAPITAL RECOVERY FACTOR GIVEN P TO FIND R	UNIFORM ANNUAL SERIES: COMPOUND AMOUNT FACTOR GIVEN R TO FIND S	UNIFORM ANNUAL SERIES: PRESENT WORTH FACTOR GIVEN R TO FIND P	N PERIODS
	(1 + I)**N	1 / (1 + I)**N	I / ((1 + I)**N - 1)	I(1 + I)**N / ((1 + I)**N - 1)	((1 + I)**N - 1) / I	((1 + I)**N - 1) / I(1 + I)**N	
1	1.2400000	.8064516	1.0000000	1.2400000	1.0000000	.8064516	1
2	1.5376000	.6503642	.4464286	.6864286	2.2400000	1.4568158	2
3	1.9066240	.5244873	.2647183	.5047183	3.7776000	1.9813031	3
4	2.3642138	.4229736	.1759255	.4159255	5.6842240	2.4042767	4
5	2.9316251	.3411077	.1242477	.3642477	8.0484378	2.7453844	5
6	3.6352151	.2750869	.0910742	.3310742	10.9800628	3.0204713	6
7	4.5076667	.2218443	.0684216	.3084216	14.6152779	3.2423156	7
8	5.5895067	.1789067	.0522932	.2922932	19.1229446	3.4212222	8
9	6.9309883	.1442796	.0404654	.2804654	24.7124513	3.5655018	9
10	8.5944255	.1163545	.0316021	.2716021	31.6434396	3.6818563	10
11	10.6570876	.0938343	.0248522	.2648522	40.2378651	3.7756906	11
12	13.2147887	.0756728	.0196483	.2596483	50.8949527	3.8513634	12
13	16.3863379	.0610264	.0155983	.2555983	64.1097414	3.9123898	13
14	20.3190590	.0492149	.0124230	.2524230	80.4960793	3.9616047	14
15	25.1956332	.0396894	.0099191	.2499191	100.8151384	4.0012941	15

24.00%

16	31.2425852	.0320076	.0079358	.2479358	126.0107716	4.0333017	16
17	38.7408056	.0258126	.0063592	.2463592	157.2533568	4.0591143	17
18	48.0385990	.0208166	.0051022	.2451022	195.9941624	4.0799309	18
19	59.5678627	.0167876	.0040978	.2440978	244.0327614	4.0967184	19
20	73.8641498	.0135384	.0032938	.2432938	303.6006241	4.1102568	20
21	91.5915457	.0109180	.0026493	.2426493	377.4647739	4.1211748	21
22	113.5735167	.0088049	.0021319	.2421319	469.0563196	4.1299797	22
23	140.8311607	.0071007	.0017164	.2417164	582.6298363	4.1370804	23
24	174.6306393	.0057264	.0013822	.2413822	723.4609971	4.1428068	24
25	216.5419927	.0046180	.0011135	.2411135	898.0916364	4.1474248	25
26	268.5120710	.0037242	.0008972	.2408972	1114.6336291	4.1511491	26
27	332.9549680	.0030034	.0007230	.2407230	1383.1457001	4.1541525	27
28	412.8641603	.0024221	.0005827	.2405827	1716.1006681	4.1565746	28
29	511.9515588	.0019533	.0004697	.2404697	2128.9648284	4.1585279	29
30	634.8199329	.0015752	.0003787	.2403787	2640.9163873	4.1601031	30
31	787.1767168	.0012704	.0003053	.2403053	3275.7363202	4.1613735	31
32	976.0991289	.0010245	.0002461	.2402461	4062.9130370	4.1623980	32
33	1210.3629198	.0008262	.0001985	.2401985	5039.0121659	4.1632242	33
34	1500.8500206	.0006663	.0001600	.2401600	6249.3750858	4.1638905	34
35	1861.0540255	.0005373	.0001290	.2401290	7750.2251063	4.1644278	35
36	2307.7069916	.0004333	.0001040	.2401040	9611.2791319	4.1648611	36
37	2861.5566696	.0003495	.0000839	.2400839	11918.9861235	4.1652106	37
38	3548.3302704	.0002818	.0000677	.2400677	14780.5427932	4.1654924	38
39	4399.9295352	.0002273	.0000546	.2400546	18328.8730635	4.1657197	39
40	5455.9126237	.0001833	.0000440	.2400440	22728.8025988	4.1659030	40

NOTE- **N IS EXPONENT N

25.00 PERCENT COMPOUND INTEREST FACTORS

N PERIODS	SINGLE PAYMENT: COMPOUND AMOUNT FACTOR GIVEN P TO FIND S	SINGLE PAYMENT: PRESENT WORTH FACTOR GIVEN S TO FIND P	UNIFORM ANNUAL SERIES: SINKING FUND FACTOR GIVEN S TO FIND R	UNIFORM ANNUAL SERIES: CAPITAL RECOVERY FACTOR GIVEN P TO FIND R	UNIFORM ANNUAL SERIES: COMPOUND AMOUNT FACTOR GIVEN R TO FIND S	UNIFORM ANNUAL SERIES: PRESENT WORTH FACTOR GIVEN R TO FIND P	N PERIODS
	(1 + I)**N	1 / (1 + I)**N	I / ((1 + I)**N - 1)	I(1 + I)**N / ((1 + I)**N - 1)	((1 + I)**N - 1) / I	((1 + I)**N - 1) / I(1 + I)**N	
1	1.2500000	.8000000	1.0000000	1.2500000	1.0000000	.8000000	1
2	1.5625000	.6400000	.4444444	.6944444	2.2500000	1.4400000	2
3	1.9531250	.5120000	.2622951	.5122951	3.8125000	1.9520000	3
4	2.4414063	.4096000	.1734417	.4234417	5.7656250	2.3616000	4
5	3.0517578	.3276800	.1218467	.3718467	8.2070313	2.6892800	5
6	3.8146973	.2621440	.0888195	.3388195	11.2587891	2.9514240	6
7	4.7683716	.2097152	.0663417	.3163417	15.0734863	3.1611392	7
8	5.9604645	.1677722	.0503985	.3003985	19.8418579	3.3289114	8
9	7.4505806	.1342177	.0387562	.2887562	25.8023224	3.4631291	9
10	9.3132257	.1073742	.0300726	.2800726	33.2529030	3.5705033	10
11	11.6415322	.0858993	.0234929	.2734929	42.5661287	3.6564026	11
12	14.5519152	.0687195	.0184476	.2684476	54.2076609	3.7251221	12
13	18.1898940	.0549756	.0145434	.2645434	68.7595761	3.7800977	13
14	22.7373675	.0439805	.0115009	.2615009	86.9494702	3.8240781	14
15	28.4217094	.0351844	.0091169	.2591169	109.6868377	3.8592625	15

25.00%

16	35.5271368	.0281475	.0072407	.2572407	138.1085472	3.8874100	16
17	44.4089210	.0225180	.0057592	.2557592	173.6356839	3.9099280	17
18	55.5111512	.0180144	.0045862	.2545862	218.0446049	3.9279424	18
19	69.3889390	.0144115	.0036556	.2536556	273.5557562	3.9423539	19
20	86.7361738	.0115292	.0029159	.2529159	342.9446952	3.9538831	20
21	108.4202172	.0092234	.0023273	.2523273	429.6808690	3.9631065	21
22	135.5252716	.0073787	.0018584	.2518584	538.1010862	3.9704852	22
23	169.4065895	.0059030	.0014845	.2514845	673.6263578	3.9763882	23
24	211.7582368	.0047224	.0011862	.2511862	843.0329473	3.9811105	24
25	264.6977960	.0037779	.0009481	.2509481	1054.7911841	3.9848884	25
26	330.8722450	.0030223	.0007579	.2507579	1319.4889801	3.9879107	26
27	413.5903063	.0024179	.0006059	.2506059	1650.3612251	3.9903286	27
28	516.9878828	.0019343	.0004845	.2504845	2063.9515314	3.9922629	28
29	646.2348536	.0015474	.0003875	.2503875	2580.9394142	3.9938103	29
30	807.7935669	.0012379	.0003099	.2503099	3227.1742678	3.9950482	30
31	1009.7419587	.0009904	.0002478	.2502478	4034.9678347	3.9960386	31
32	1262.1774484	.0007923	.0001982	.2501982	5044.7097934	3.9968309	32
33	1577.7218104	.0006338	.0001586	.2501586	6306.8872418	3.9974647	33
34	1972.1522631	.0005071	.0001268	.2501268	7884.6090522	3.9979718	34
35	2465.1903288	.0004056	.0001015	.2501015	9856.7613153	3.9983774	35
36	3081.4879110	.0003245	.0000812	.2500812	12321.9516441	3.9987019	36
37	3851.8598888	.0002596	.0000649	.2500649	15403.4395551	3.9989615	37
38	4814.8248610	.0002077	.0000519	.2500519	19255.2994439	3.9991692	38
39	6018.5310762	.0001662	.0000415	.2500415	24070.1243048	3.9993354	39
40	7523.1638453	.0001329	.0000332	.2500332	30088.6553811	3.9994683	40

NOTE- **N IS EXPONENT N

APPENDIX B

Selected References

The following list of references is meant to be neither comprehensive nor exclusive, nor can it be considered a bibliography for the material presented in this book. The books listed here are simply believed to be among the best published on the various topics, and will give the engineer a good overall reference library on the most important aspects of cost engineering and economic analysis.

Engineering Economics

Grant, E. L. *Principles of Engineering Economy*. Ronald, New York, 1964.

Thuesen, H. G. *Engineering Economy*. Prentice-Hall, Englewood Cliffs, N. J., 1964.

Barish, N. M. *Economic Analysis for Engineering and Managerial Decision-Making*. McGraw-Hill, New York, 1962.

Smith, G. W. *Engineering Economy: Analysis of Capital Expenditures*. Iowa State University Press, Ames, Iowa, 1968.

Investment Evaluation and Financial Analysis

Hackney, J. W. *Control and Management of Capital Projects*, Wiley, New York, 1965.

Porterfield, J. T. S. *Investment Decisions and Capital Costs*. Prentice-Hall, Englewood Cliffs, N. J., 1965.

Helfert, E. *Techniques of Financial Analysis*. Irwin, Homewood, Ill., 1967.

Cohen, J. B. and S. M. Robbins. *The Financial Manager*. Harper, New York, 1966.

Bierman, H. J. and S. Smidt. *The Capital Budgeting Decision*. Macmillan, New York, 1966.

Gillis, F. E. *Managerial Economics: Decision Making under Uncertainty for Business and Engineering*. Addison-Wesley, Reading, Mass., 1969.

Weston, J. F. and Brigham, E. F. *Essentials of Managerial Finance*. Holt, New York, 1968.

Fleischer, G. A. *Capital Allocation Theory: The Study of Investment Decisions*. Appleton-Century-Crofts, New York, 1969.

Depreciation

Grant, E. L. and P. T. Norton. *Depreciation*. Ronald, New York, 1955.

Marston, Winfrey, and Hemstead. *Engineering Valuation and Depreciation*. McGraw-Hill, New York, 1953.

U.S. Treasury Department, Internal Revenue Service. *Depreciation Guidelines and Rules*.

Cost and Profit Analysis

Bauman, H. C. *Fundamentals of Cost Engineering in the Chemical Industry*. Van Nostrand Reinhold, New York, 1964.

Chilton, C. H. *Cost Engineering in the Process Industries*. McGraw-Hill, New York, 1960.

Dearden, J. *Cost and Budget Analysis*. Prentice-Hall, Englewood Cliffs, N. J., 1962.

Gardner, F. V. *Profit Management and Control*. McGraw-Hill, New York, 1955.

Park, W. R. *The Strategy of Contracting for Profit*. Prentice-Hall, Englewood Cliffs, N. J., 1966.

Jelen, F. C. *Cost and Optimization Engineering*. McGraw-Hill, New York, 1970.

Popper, H. *Modern Cost Engineering Techniques*. McGraw-Hill, New York, 1970.

Economic Forecasting and Models

Butler, W. F. and R. A. Kavesh, Eds. *How Business Economists Forecast*. Prentice-Hall, Englewood Cliffs, N. J., 1966.

Christ, C. F. *Econometric Models and Methods*. Wiley, New York, 1966.

Bean, L. H. *The Art of Forecasting*. Random House, New York, 1969.

Silk, L. S. and L. M. Curley. *Primer on Business Forecasting with a Guide to Sources of Business Data*. Random House, New York, 1970.

Index